Strategies to Modify the Drug Release from Pharmaceutical Systems

Related titles

Rathbone, Hadgraft, Roberts and Lane,
Modified-Release Drug Delivery Technology,
Second Edition, CRC Press, 2008, 9781420044355

Woodhead Publishing Series in Biomedicine:
Number 85

Strategies to Modify the Drug Release from Pharmaceutical Systems

Edited by

Marcos Luciano Bruschi

ELSEVIER

AMSTERDAM • BOSTON • CAMBRIDGE • HEIDELBERG
LONDON • NEW YORK • OXFORD • PARIS • SAN DIEGO
SAN FRANCISCO • SINGAPORE • SYDNEY • TOKYO
Woodhead Publishing is an imprint of Elsevier

WP
WOODHEAD
PUBLISHING

Woodhead Publishing Limited is an imprint of Elsevier
80 High Street, Sawston, Cambridge, CB22 3HJ, UK
225 Wyman Street, Waltham, MA 02451, USA
Langford Lane, Kidlington, OX5 1GB, UK

ISBN: 978-0-08-100092-2 (print)
ISBN: 978-0-08-100112-7 (online)

British Library Cataloguing in Publication Data
A catalogue record for this book is available from the British Library

Library of Congress Cataloging-in-Publication Data
A catalog record for this book is available from the Library of Congress

Library of Congress Number: 2015942326

For Information on all Woodhead Publishing publications
visit our website at http://store.elsevier.com/

Contents

Preface

The research in therapeutic systems has increased quickly with the development of new materials and new mathematical approaches, and further application of physicochemical tools. Moreover, the great development of biotechnology and the understanding of physiological mechanisms have enabled the possibility of designing dosage forms that are each more specialized. This is increasing the number of strategies that can be applied to develop systems with enhanced properties in relationship to modifying and controlling the delivery of active agents.

This book, "Strategies to Modify the Drug Release from Pharmaceutical Systems," emerged with the aim of providing a comprehensive reference on drug delivery and strategies to modify the drug release from dosage forms, covering all main topics in the field. From the earliest dosage forms to modern drug delivery systems has come great development and growth of acquired knowledge. Thus, this textbook addresses principles, systems, applications, and advances in the field. It is principally a textbook and a reference source of strategies to modify the drug release. Moreover, the characterization, mathematical and physicochemical models, applications, and systems are discussed.

Some students and researchers may consider this book an excellent starting point to study the field, while others may find it a source of principles, systems, applications, and advances in the field of drug delivery systems addressed in the same book. It is designed for individuals of diverse backgrounds who are interested in the field of drug delivery. The mathematical and physicochemical principles are highlighted and related with the strategies aiming to discuss and understand the drug release and its possible modifications. It is a book involving the principles, systems, mechanisms, and the state of the art in just one volume.

The book is divided into six chapters, according to its sixfold purpose: (I) General considerations, (II) Modification of drug release, (III) Classification of therapeutic systems for drug delivery, (IV) Main mechanisms to control the drug release, (V) Mathematical and physicochemical models of drug release, and (VI) Drug delivery systems. Each chapter is devoted to a specific group of concepts and definitions, and contains contributions from established or emerging researchers around the world who have been working in this field.

Through many years of studying and lecturing on this theme for undergraduate and graduate students, I have observed the need to have in one book the discussion about the main strategies to modify the drug delivery and the fundamentals about the theme. I sincerely hope that this textbook will help pharmaceutical students and scientists to understand the fundamentals and main points involving the strategies to modify the drug release. The knowledge about drug delivery systems will enable a more specialized understanding, research and development of drug delivery systems, contributing to patient compliance and enhanced therapy.

This book is the outcome of more than 15 years of experience in pharmaceutical compounding and drug delivery. I am indebted to many people for assisting and guiding my knowledge in the field throughout this period, from my experience at the School of Pharmacy of the State University of Maringá (UEM), to the pharmacy faculties of the State University of São Paulo (UNESP) and of the University of São Paulo (USP), to the periods of PhD studies and postdoctoral studies at the School of Pharmacy of Queen's University of Belfast (QUB). I wish to thank Professor Maria Palmira Daflon Gremião and Professor Raul Cesar Evangelista (in memoriam) (pharmaceutical sciences faculty of UNESP—Araraquara—Brazil) and Professor Selma Lucy Franco (School of Pharmacy—UEM—Maringá—Brazil) for introducing and guiding me through the vast array of drug delivery, every time with enthusiasm and patience. I would like to thank Professor Osvaldo de Freitas and Professor Elza Helena Guimarães Lara (pharmaceutical sciences faculty of Ribeirão Preto—USP—Brazil) for their important contributions to my knowledge about pharmaceutical compounding. Special thanks to Professor Heitor Panzeri (in memoriam) (dentistry faculty of Ribeirão Preto—USP—Brazil) for the teachings in dentistry and dental material fields. I very much thank Professor David S. Jones (School of Pharmacy—QUB—Belfast, Northern Ireland, UK) for the opportunity to expand my knowledge in drug delivery and for helping me with my questions with patience and enthusiasm, every time. Very special thanks for the contributions of the students to whom I have had the pleasure of teaching pharmaceutics, physical pharmacy, compounding, pharmaceutical technology, and drug delivery over several years.

Finally, I would like to thank my wife, Michelle, our children, Gabriel and Beatriz, and my parents, Marco and Cleusa. Without their support, patience, kindness, love, and encouragement this book would not have been completed. This work is dedicated to all my family.

Marcos L. Bruschi

General considerations

1

1.1 Pharmaceutics: Safety, quality, and efficacy

The origin of modern humans is from Africa and dates from approximately 100,000 years ago (Pena, 2002). Despite being relatively new, *Homo sapiens* possess a complex organism that originated from a unique cell. When the egg is fertilized, it divides into two cells that divide again, which result in four, and so on. Besides the division, these basic units of organisms differentiate and originate tissues, organs, and systems. Their equilibrium and homeostasis are very important to the body's global health. However, diseases and disequilibrium are very common, and the constant evolution of science, aiming for cures and well-being, conduced to the utilization of the different kingdoms of nature.

Humans, with their intelligence, search constantly for new drugs and materials. The World Health Organization defines a drug as "any substance that is used to modify or explore physiological systems or pathological states for the benefit of the patient." It is an active agent intended for use in diagnosis, mitigation, treatment, cure, or prevention (FDA, 1938).

The diversity of drug actions and effects on the body is very important and enables their selective use in the treatment of a range of common and rare conditions involving virtually every body organ, tissue, and cell (Allen, Popovich, & Ansel, 2011). However, adjuvants are necessary to make possible the administration of an active agent. Thus, the concept of formulation and dosage form is introduced when we observe that medicines rarely are composed just drugs.

Medicine, or formulated preparation, is defined as a pharmaceutical formulation containing one or more drugs, to be used in diagnosis, prevention and/or cure of diseases and their symptoms or the correction/modification of organic functions, either in human being or in animals. It is a means of administering active agents into the body in a safe, efficient, reproducible, and convenient manner.

Until the mid-nineteenth century, the production of drugs was essentially handmade, consisting of drugs from plant, mineral or animal origin. During the early twentieth century, scientific institutions focused on the implementation of the research and production of medicines, vaccines and serums. The beginning of fermentation and chemical synthesis as a technological process started in the 1930s. Between 1940 and 1960, the global pharmaceutical industry emerged with the launch of new products and changes in marketing strategies, influencing the prescription.

The thalidomide disaster led to more rigorous registration policies, clinical trials, and quality control of products, seeking the safety, efficacy, and quality of medicines. Thus, the accuracy required in the approval of drugs, demanding the accomplishing of expensive trials in humans, together with the gradual occupation of the vast majority

Strategies to Modify the Drug Release from Pharmaceutical Systems. http://dx.doi.org/10.1016/B978-0-08-100092-2.00001-1

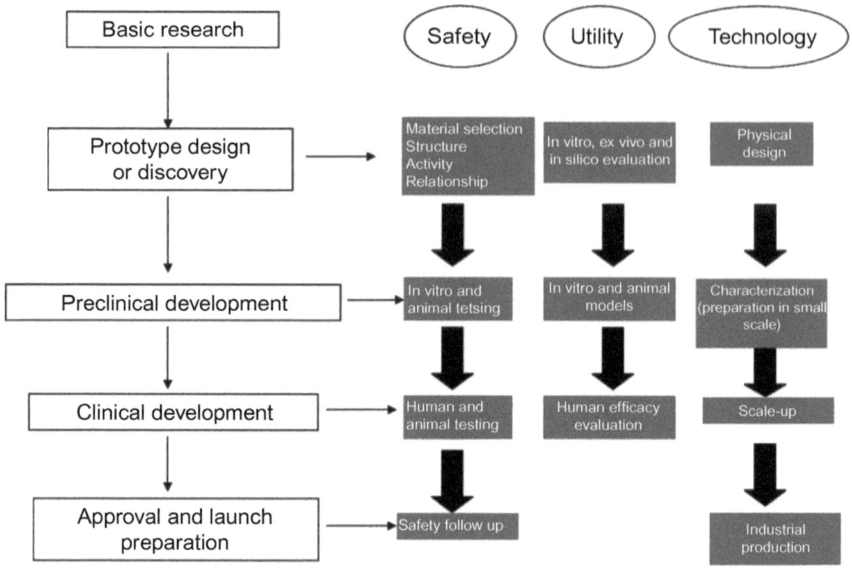

Figure 1.1 Generalized description of steps that must be successfully completed at different points and in different dimensions along the process of development and production of a medicine.

of the market's niche by innovations over time, have significantly raised the cost of developing a new drug. Figure 1.1 shows the steps to the process of medicine development. Many activities must be successfully completed at different points and in different dimensions along the critical path. They are performed for every product, and many of them are highly complex (FDA, 2004).

Pharmaceutical sciences are involved with the design, development, production, and use of medicines. Pharmaceutics is a scientific discipline of pharmacy that is about the conversion of a drug into a medicine suitable for administration by or to patients. The understanding of pharmaceutical preparation is very important, and the study of drug formulations and their design, manufacture, and delivery into the body is related to pharmaceutics (Florence & Attwood, 2006). Physical properties, dosage form design, and the manufacture of these medicines on both a small (compounding) and a large (pharmaceutical technology) scale are included in pharmaceutics (Aulton, 2001; Allen et al., 2011).

In the design of dosage forms, the physicochemical properties of the drug and biopharmaceutical and therapeutic considerations are very important:

What will be the administration route?

What is the disease to be treated?

What is the dose frequency for the drug?

The answers to these questions are very important to achieve the most suitable therapy. The medicine may vary from relatively simple solutions to complex drug delivery systems, using appropriate additives, excipients, or materials in the preparations. The aim is to achieve a better, more predictable therapeutic response to a drug included in a formulation, capable of being produced in large scale with reproducible quality.

To achieve this, microbiological, physical and chemical stability; precision of drug dose; packaging; labeling; and acceptability to users (prescriber and patient) are fundamental points that must be considered. Moreover, the natural variations between patients cannot interfere. This is difficult to achieve, but new knowledge, technologies and the vast array of drugs available today make it achievable. Nowadays, new active agents originated from animal, mineral or plant sources; chemical synthesis; molecular modification; biotechnology; or microbial technology are very important contributions for new medicines. Moreover, the evolution of computational technology has assisted the drug discovery, contributing to produce data banks and libraries of chemical compounds and methods of screening for potential biologic activity (Allen et al., 2011), and the obtaining of new dosage forms as well.

In this sense, for the development of new medicine, the first step is to know the chemical, physical, and biological characteristics of the active agent. The pharmacology of the drug and toxicological features must be determined. The drug's site and rate of absorption, its pattern of distribution and concentration within the body, its duration of action, and the method and rate of its elimination or excretion must be studied (Allen et al., 2011).

Biopharmacy is a branch of pharmaceutical research that studies the physiological and pharmaceutical factors influencing drug release from the dosage form, its absorption by the body and bioavailability. The drug metabolism and the activity of any of its metabolites must be obtained. The short- and long-term effects of the active agent on body cells, tissues, and organs must be studied, as well as the drug's ability to pass to a nursing baby through the mother's breast milk, the hazardous effects of the drug on the fetus, and the complexation with plasmatic proteins.

Governmental agencies view the security and efficacy characteristics of new chemical entities each time more closely. The understanding of toxicological factors and metabolism of active agents is dependent on the patient age, drug distribution, genetic factors, time of exposition, and potential for teratogenic, mutagenic and embryotoxic effects. It is very important to investigate carefully to minimize possible toxic reactions, showing the safety and efficacy of new active agents.

The study of approved drugs conduces to the development of dosage forms with more efficacious delivery to the appropriate site, optimizing bioavailability, minimizing toxicity and side effects, and improving stability (Maurin, Hussain, & Dittert, 2006).

1.1.1 Routes of drug administration

Drugs can be introduced into the body by many routes, such as enteric (oral, *peroral*, rectal), parenteral (intravascular, intramuscular, subcutaneous, and inhalation administration) or topical (skin and mucosal membranes), among others, and must be determined considering the varying of individual ages (e.g., neonates, children, adults, geriatrics), weights and states of illness. Each route has specific purposes, advantages, and disadvantages (Table 1.1) (Allen et al., 2011; Aldridge, 2010).

The most convenient and acceptable method of receiving medication is by mouth (oral). Most medicines taken by mouth are intended to be swallowed, and are referred to as *peroral* medicines. Others, known as *sublingual*, are specifically for dissolving

Table 1.1 Main routes of medicine administration: advantages and disadvantages

Route	Advantages	Disadvantages
Oral	Cheap, easy, no special equipment. Acceptable to most people. Suitable for self-medication	May be compromised by irritant effects/presence of food. Enzyme action may limit effectiveness
Sublingual	Drug absorption through buccal or sublingual mucosa avoids gut enzymes. Rapid action	Taste of drug may be a problem
Transdermal	Easy to use. Long action can be achieved. Avoids adverse effects of gastrointestinal tract enzymes	Relatively high cost. Drug may build up in skin so that action continues when patch is removed
Inhalation	Rapid action (inhaled anesthetics). Limits systemic absorption. Avoids gut enzymes	Needs specialized drug delivery system. Loss of dose—patient swallows most of the drug. Technique needs to be taught
Intranasal	Similar to inhalation	May irritate nasal mucosa. Needs special drug delivery system. Absorption may vary
Subcutaneous	Rapid absorption. Bypasses gastrointestinal tract. Patients may be taught to use this method	Absorption may be too rapid
Intramuscular	Good absorption. Bypasses gastrointestinal tract	Local irritancy. May be painful. Hazard of nerve damage. Skill involved
Intravenous	Rapid action can control rate of administration. Suitable for large volumes and drugs that would cause intramuscular irritation	Relatively high cost. Skill involved. Extravasation risk. Specialist drug delivery system needed
Rectal	Suitable for drugs that may irritate the upper gut. Fairly rapid action	May not be acceptable to some people. Variable absorption
Vaginal	Suitable for drugs that may irritate the upper gut. Fairly rapid action	Variable absorption. Needs special drug delivery system
Ophthalmic	Suitable for the external eye. Rapid action and systemic absorption	Variable absorption. Needs special drug delivery system. May irritate ocular mucosa

under the tongue; some, known as *buccal*, are for holding against the mucous membranes of the cheek (Aldridge, 2010; Mignani et al., 2013). Intraperiodontail pocket is a particular type of buccal administration by which the medicine is administered into the pocket formed during the inflammatory and infectious processes of periodontal disease (Medlicott, Rathbone, Tucker, & Holborow, 1994; Bruschi et al., 2007). Many drugs can be rectally administered in the cases of people who cannot take a drug orally because they have nausea, cannot swallow, or have restrictions on eating, as is required after many surgical procedures. Parenteral administration is a very important route to avoid passing through the gastrointestinal tract. The drug directly reaches the blood and, sometimes, the absorption phase is bypassed (intravascular), producing a rapid drug effect. Topical administration is the route by which the active agent is administered on the skin and mucous membrane for the local or systemic action.

Basically, the accessibility of active agent molecules to their respective target site and, consequently, the drug treatment is strongly dependent on the way the medicine is administered. Therefore, the route of administration has a profound effect upon both the speed and the efficiency with which the drug acts. Each active agent possesses different absorption and distribution profiles. These determine the rate at which an active agent reaches it site of action and, consequently, its response.

It is evident that no single route of administration of a drug is ideal for all active agents in all circumstances (Chugh, Seth, Rana, & Gupta, 2012). This is determined primarily by the properties of the drug (e.g., ionization, hydrophilicity, or hydrophobicity) and by the therapeutic aims (e.g., the need for rapid onset of action, the desirability of long-term treatment, or restriction of delivery to a local site).

For each administration route, there are appropriate dosage forms, such as tablets, capsules, injections, suppositories, ointments, aerosols, suspensions, emulsions, and others (Allen et al., 2011). These formulations are designed to contain a determined content of active agent for appropriate, easy, and accurate dose administration. Each formulation is unique and, in addition to the therapeutic agent, a number of non-therapeutic or pharmaceutical adjuvants are present. Materials as fillers, thickeners, solvents, suspending agents, tablet coatings and disintegrants, penetration enhancers, stabilizing agents, antimicrobial preservatives, flavors, colorants, and sweeteners are responsible for producing the form and particular characteristics.

The addition of a cosolvent, pH, a ligand for complexation, and the presence of surfactants help the solubilization of the active agent at the site of delivery. However, these techniques are limited. For example, the rapid intravenous administration of an active agent solubilized can result in rapid dilution in an environment in which it is insoluble, and an incomplete availability and a delayed response due to the formation of particulate matter within the vascular system are produced. On the other hand, this poor aqueous solubility may be desirable for the drug to sustain the effect after administration. When the drug is poorly soluble in water and it is adsorbed throughout the gastrointestinal tract, the oral sustained release may be taking place. After intramuscular administration of an active agent in suspension that is poorly soluble in water, the parenteral sustained release can be achieved. This kind of release is a function of its low solubility under physiologic conditions, its precipitation, or the formation of a reservoir or depot from an oil-containing dosage form (Maurin et al., 2006).

1.1.2 Biopharmaceutical aspects

The design of an efficacious dosage form is dependent on the understanding of the pathways of the physiologic disposition of the drug. The drug administration, release, absorption, first-pass metabolism, metabolism and formation of the active agent can result in incomplete bioavailability. The investigation of these parameters is very important for selecting and developing an optimum dosage form (Maurin et al., 2006).

Biopharmaceutics investigates the physicochemical properties that affect the extent and speed of absorption or release of a drug from a medicine or dosage form, considering the route of administration and drug distribution, metabolism, retention, and excretion. It is the study of the influence of formulation on the therapeutic efficacy of an active agent. The relationship between the nature and intensity of biological effects is determined by the type of drug, physical state, particle size, surface area, presence of additional substances, nature of supply form in which the drug is administered and methods of pharmaceutical production to formulate the drug (Florence & Attwood, 2006). Therefore, biopharmaceutical investigation is of fundamental importance in the development of a medicine.

The intravenous route enables a drug to be introduced directly into the bloodstream. However, when a systemic action is required, all other routes of administration involve the absorption of the drug into the blood from the route of administration.

Absorption is defined by the rate and extent to which the active agent leaves the place of administration and reaches blood concentration. It is dependent on the anatomy, and physiology of the administration site, the physiochemical properties of the drug and the type of formulation. Passive diffusion means that many drugs are absorbed in the gastrointestinal tract, where the rate of absorption is proportional to the drug concentration gradient across the barrier. Other drugs are absorbed by a combination of passive and active transport by proteins that can increase and/or decrease absorption, depending on their location and whether they are influx or efflux transporters (Anderson & Saneto, 2012). Drug absorption by passive diffusion is influenced by its oil-aqueous partition coefficient, but there is not a strict relationship between the partition coefficient of a molecule and its ability to be absorbed, due to the complex nature of the absorption process. Many factors can affect drug absorption across membranes, such as the ionic and/or polar characteristics of the drug and/or membrane as well as the site and capacity of carrier-mediated absorption or efflux systems (Maurin et al., 2006).

Once in the bloodstream, the drug partitions between the plasma and the erythrocytes. The portion in plasma can complex with plasma proteins, mainly albumin, or disperse in the plasma water. The latter can pass out of the plasma through the capillary endothelium and reach other body fluids and tissues and, hence, the place of action. There is a dynamic equilibrium between the amount of the drug in the blood and at the site of action. This distribution depends on the physicochemical properties of the active agent, mainly its lipophilicity.

The determination of unbound drug concentration in plasma gives a better estimate of the drug at the site of action, because direct access is often difficult. A sensitive and complex method is necessary to determine the concentration of the bound and the unbound drug within the plasma. For clinical purposes, the total drug concentration within the plasma is usually measured, and the plasma protein binding is a crucial parameter to investigate with regard to the therapeutic effect of an active molecule.

Therefore, absorption and extent of distribution are important factors that determine the plasma concentration of the drug. However, the rate of elimination of the drug from the body is also important, and can occur either for the molecule unchanged or for its metabolites originating from enzymatic cleaving or biochemical transformation. The study and characterization of the course of drug absorption, distribution, metabolism and elimination is defined as pharmacokinetics, and it is used in the clinical setting to enhance the safe and effective therapeutic management of individual patients.

Considering the routes of medicine administration, it is possible to relate the pharmacokinetic parameters, considering the body as a model of just one compartment (Figure 1.2).

When the active substance is intravenously administered, it reaches the systemic circulation directly and is totally (100%) bioavailable. However, if another route is used to administer the drug, there is no guarantee that the whole dose will reach the systemic circulation intact. In this case, the amount of the drug that reaches the systemic circulation in the unchanged form is a fraction of the administered dose (bioavailable dose). Thus, bioavailability refers to the amount of active substance that is absorbed by the organism in a certain amount of time and that arrives at the target site of action. In these terms, it is assumed that the intact drug is the therapeutically active form. However, this definition cannot be considered when therapeutic action depends on drug conversion into a therapeutically active form prior to or upon reaching systemic circulation (prodrugs). Moreover, in the context of bioavailability, the term systemic circulation refers primarily to venous blood and the arterial blood, which carries the intact blood to the tissues. The hepatic portal vein, which carries blood from the gastrointestinal tract to the liver in the absorption phase, is not considered.

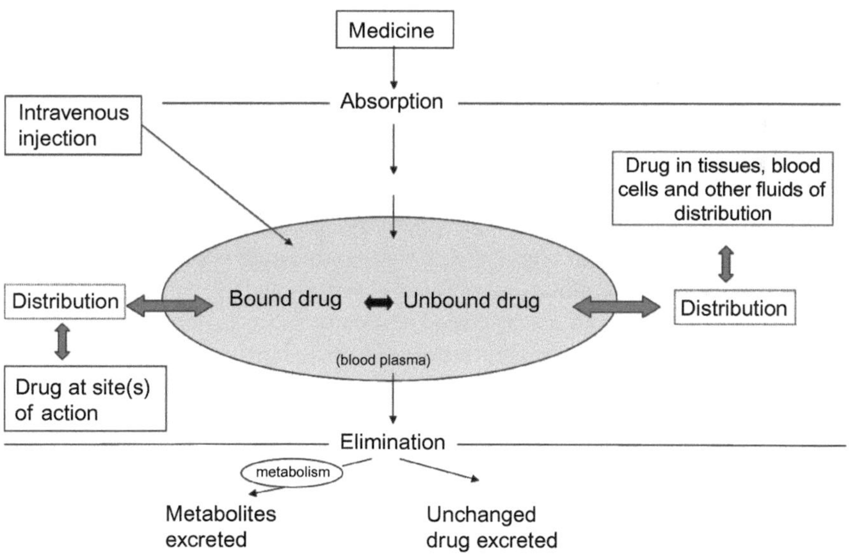

Figure 1.2 Diagrammatic representation of the absorption, distribution, and elimination of an active agent.

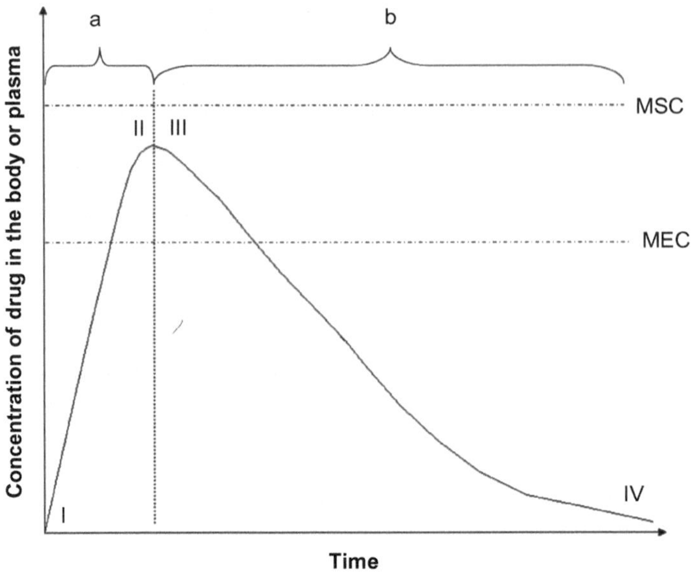

Figure 1.3 Body or plasma concentration-time curve of a drug obtained following the peroral administration of a dosage form in the body that confers one-compartment, open-model characteristics on the body: (a) Absorption phase, (b) elimination phase, (I-II) rate of drug absorption higher than rate of drug elimination, (III-IV) rate of drug absorption less than rate of drug elimination, (II-III) rate of drug input equal to rate of drug output, (MSC) maximum safe concentration, (MEC) minimum effective concentration.

Therefore, any effect in either the release of the drug from the dosage form, its dissolution into the gastrointestinal fluids, its permeation through and stability in the gastrointestinal barrier or its stability in the hepatic portal circulation will influence the bioavailability exhibited by that drug from the dosage form in which it was administered (Ashford, 2001). The plasma concentration-time curve after a single oral dose of an active substance as well as the parameters associated with a therapeutic effect may be visualized in Figure 1.3.

Considering the one-compartment, open model of bioavailability of a drug (Figure 1.3), the dosage form administered by peroral route starts to release the drug, which passes across the cellular membranes (gastrointestinal barrier). The result of these processes is the rate of absorption. The rate of input is dependent on time and drug concentration, and the apparent absorption rate constant exists. This constant gives the fraction of active agent that enters the compartment (body) per unit of time.

In addition, the rate of drug absorption will decrease gradually with time, because the drug concentration at the absorption site will decrease with time as well. Chemical degradation and movement of the active agent away from the absorption site will also contribute to the gradual decrease in the effective drug concentration (Proudfoot & Collett, 2001). The magnitude of drug elimination is dependent on the amount of active agent in the compartment (body) at the specified time, and the rate of this process follows the first-order model. Therefore, drug absorption is decreasing and drug elimination is increasing with time, and the total concentration of the drug in the body will reach a peak value and then begin to fall with time (Figure 1.3). When there is no equilibrium between the absorption and elimination rate of the drug, or the amount of the drug administered is very high, the peak concentration achieved in the body or plasma reaches toxic concentration and the patient can exhibit side effects.

Different formulations, dosage forms and/or routes of administration can exhibit variability in the drug bioavailability. Sometimes the plasma concentration of the drug is too high, causing side effects; other times it is too low, and the drug is ineffective (Figure 1.4). The drug must be released from the formulation in the proper amount and in such a manner that the onset and duration of the drug's action are that which are desired.

The dosage form needs to be correctly developed to provide a proper administration for the patient to receive maximum benefit. Therefore, the amount of medication, the time intervals, and the indicated duration to achieve the desired therapeutic outcomes are very important. The conception of therapeutic regimen is fundamental and the objective is to optimize administration of the medicine and control the therapeutic effect of the drug, considering the *in vivo* variations of the drug release. When this aim is accomplished, the maximum bioavailability of the drug is reached, and there is control of release and absorption.

The conception of dosage regimen is fundamental, and the biological half-life of the drug (time required for the body to eliminate 50% of the drug that it contained)

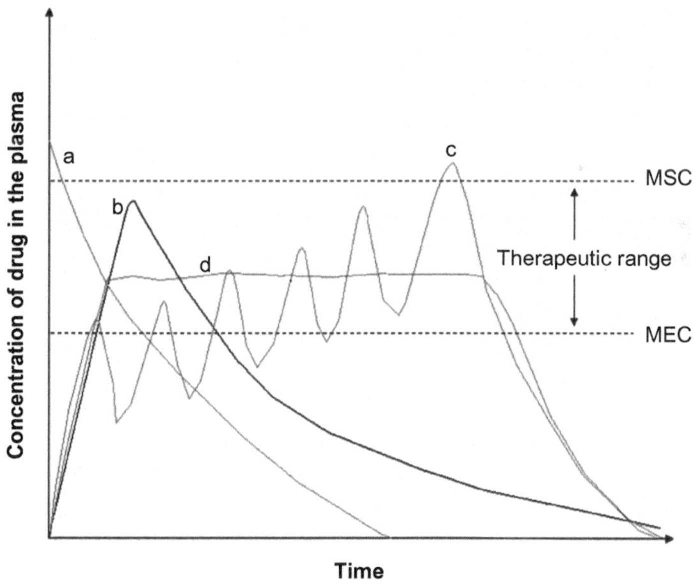

Figure 1.4 Plasma concentration-time curves of a drug obtained following the different administrations: (a) Intravenous route, (b) peroral route of single dose, (c) peroral route multiple/repeated doses, (d) peroral route of single dose with prolonged therapeutic plasma level, (MSC) maximum safe concentration, (MEC) minimum effective concentration.

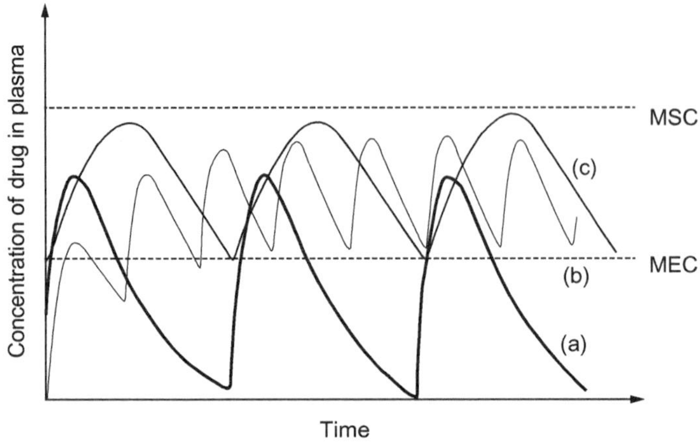

Figure 1.5 Diagrammatic representation of different dosage regimens for peroral administration of (a) equal doses of an active agent at intervals of time that allow the complete elimination of the previous dose, (b) equal doses of an active agent at fixed intervals of time equal to the biological half-life, (c) fluctuation of concentration of the drug at steady state after repetitive administration of equal doses at fixed intervals of time.

is important information. When a single dose of an active agent with a biological half-life of 8 h is administered, the peak blood concentration reached will fall by 50% every 8 h until the entire agent has been eliminated. It is very important to understand the relationship between a drug's number of half-lives elapsed and its percentage of elimination from the body.

Thus, the amount of drug administered, half-life and elimination are very important information to develop a correct dosage regimen. Figure 1.5 displays different dosage regimens and their consequences.

Therefore, the amount of drug absorbed following each dose administered and the dosage time interval in relation to the biological half-life of the drug, or both, determine the magnitude of the fluctuations between the maximum and minimum steady-state amounts of active agent in the body. This is a very important consideration for any drug with a narrow therapeutic range. To ensure that a multiple-dosage regimen provides steady-state concentrations of active agent in the body, showing clinical efficacy and safety, proper selection of dose size, and dosage time is fundamental. The first or loading dose is a large single dose of the drug that may be administered initially in order to achieve peak concentration, reducing the time required for onset of the full therapeutic drug effect. Thereafter, smaller, equal doses are administered repetitively at suitable fixed intervals so as to maintain the plasma concentrations of the drug at the maximum, minimum and average state levels that provide the patient with the full therapeutic benefit (Proudfoot & Collett, 2001; Gennaro, 2000).

The smaller proportion of the active agent being eliminated from the body over each fixed dosage time interval when the biological half-life of the drug is increased leads to the great degree of drug accumulation. The maintenance of the steady-state amount of the drug in the body may be influenced by problems on the elimination rate constant; those vary from patient to patient, and are influenced by whether the patient has normal or impaired renal function, for example.

Another very common problem is the overnight no-dose period, where the patient administers the doses at uniform time intervals during the day but not at night. This multiple-dosage regimen will give the patient periods of undisturbed sleep, but cause problems in maintaining therapeutic steady-state plasma concentrations of the drug in the body. It is not comfortable for the patient to wake up during the night to administer one or more medicine doses, and the default of overnight dose periods of 8–12 h could result in substantial decreases in the amount of the drug in the blood and body, particularly for drugs having biological half-lives that are relatively short compared to the overnight no-dose period (Proudfoot & Collett, 2001; Panchagnula & Thomas, 2000). Problems can arise with regard to maintaining therapeutic drug levels in patients, and the adhesion of the patient to therapy is prejudiced. Therefore, it is very important to administer the doses at the predetermined time intervals over 24 h to ensure the maintenance of therapeutic drug levels appropriate to the therapy of a disease. This aim can be achieved by the development of appropriate dosage forms that have the characteristics of modifying and controlling drug delivery. Once the physicochemical and biopharmaceutical properties of the drug are determined and the desired plasma or site concentration profile is defined, the pharmaceutical scientist can select and develop an efficacious system by utilizing a formulation approach, a prodrug approach, a device approach or an alternative administration route approach.

1.2 Conventional and alternative therapeutic regimens

Proper drug selection and effective dosage form are requirements for an effective therapeutic outcome. Active agents are administered into the body by different routes, and each one has specific purposes, advantages, and disadvantages. The route of administration determines the accessibility of drug molecules to their respective target site, having an important effect upon both the speed and the efficiency with which they act. Absorption and distribution are involved with the rate at which a drug reaches its site of action, and no single route of administration of a drug is ideal for all drugs in all circumstances. Moreover, the dosage form is determinant in this process, considering the physicochemical characteristics of the active agent.

In this context, the amount administered and the solubility of drug are very important. Depending on the route of administration (e.g., peroral), the absorption step can limit the rate of the drug that reaches the site of action. A drug is highly permeable when the extent of absorption is greater than 90% of the dose administered.

From pharmacokinetic studies, drugs are classified into four classes. Drugs of Class I dissolve rapidly when presented in immediate-release dosage forms and are rapidly transported across the body's barriers, displaying good availability. Class II is formed by drugs in which the dissolution rate is liable to be a rate-limiting step, making it possible to generate a strong correlation between *in vitro* dissolution and *in vivo* absorption. This class of drugs should be amenable to formulation approaches to improve the dissolution rate and, hence, oral bioavailability (Collett & Moreton, 2001). Class III drugs dissolve rapidly but are poorly permeable, so it is very important that the dosage forms release them rapidly to maximize the amount of time for active agents. Finally, Class IV drugs are poorly soluble and poorly permeable, and are liable to have poor oral bioavailability or an oral absorption so low that the drugs cannot be given by the oral route. Preparing prodrugs of these compounds or finding an alternative route of delivery are approaches that must be adopted to significantly improve their absorption into systemic circulation (Maurin et al., 2006).

Over the last century, drug discovery displayed a great evolution, with scientific researchers focusing on vegetables, chemical synthesis, and biotechnological processes. This evolution contributed to improving the therapeutics and the success of a treatment, requiring the correct diagnosis and the choice of an appropriate active agent as well as an exact plan of administration compatible with its biopharmaceutical characteristics. While new and more powerful drugs are being developed, great attention has been given to methods by which these active agents are being administered.

An acceptable therapeutic concentration of a drug at the site(s) of action attained immediately, and then maintained constant for the desired duration of treatment, is fundamental to the ideal therapeutic regimen. To accomplish these requirements, the conventional dosage forms are administered repetitively, and the drug, in a form that is therapeutically active, is released and quickly available to act locally or to be absorbed.

As previously mentioned, the concentration of a drug in the plasma or at the site of action fluctuates over successive dosing intervals. These fluctuations can lead to the patient being over- or under-medicated for time intervals. Moreover, some active agents display short biological half-lives, and the dosage regimen is composed of doses in smaller time intervals. Forgotten doses and the overnight period without dose conduce to a lack of patient compliance and therapeutic inefficiency or failure.

Conventional therapeutic regimens are characterized by the use of large amounts of a drug, with a great fraction being excreted without exerting therapeutic effects. Thus, it is necessary to extend the drug release, taking control of it considering the time and site of action.

Alternative therapeutic regimens have been proposed to provide fast relief of symptoms and longer action of drugs, protecting the patient for more time. Moreover, patient compliance to therapy increases. Continuous intravenous infusion, single doses adjusted in frequency and amount, physiological methods (reduction of elimination velocity, metabolism or absorption) and prodrugs are examples of alternative therapeutic regimens. However, treatments along these lines are far from an ideal therapy for the patient and, generally, are characterized by very long researches with high costs.

The pharmaceutical industries invest about $40 billion in research and development of new medicines, representing more than 20% of all their revenue. In the last decade, in the United States, $800 million and 15 years of research were spent on the development of just one innovative medicine (Dimasi, Hansen, & Grabowski, 2003). Therefore, over the years, an enormous amount of studies have focused on the development of dosage forms, using strategies to control the drug release in specific sites of the body in order to improve availability, reduce adverse effects, and increase drug efficacy.

The dosage forms are each time more specialized, constituting true systems that should display characteristics to release the active agent at a rate that perfectly matches the real need *in vivo* for the duration of the therapy, and to deliver the drug exclusively to its target site.

References

Aldridge, M. (2010). Miscellaneous routes of medication administration. In P. Jevon, E. Payne, D. Higgins, & R. Endacott (Eds.), *Medicines management: A guide for nurses* (pp. 239–261). Oxford: Wiley-Blackwell.

Allen, L. V., Jr., Popovich, N. G., & Ansel, H. C. (2011). *Ansel's pharmaceutical dosage forms and drug delivery*. Baltimore-Philadelphia: Lippincott Williams & Wilkins.

Anderson, G. D., & Saneto, R. P. (2012). Current oral and non-oral routes of antiepileptic drug delivery. *Advanced Drug Delivery Reviews*, *64*, 911–918.

Ashford, M. (2001). Introduction to biopharmaceutics. In M. E. Aulton (Ed.), *Pharmaceutics: The science of dosage form design* (2nd ed., pp. 213–216). London: Churchill Livingstone.

Aulton, M. E. (Ed.) (2001). *Pharmaceutics: The science of dosage form design*. New York: Churchill Livingstone.

Bruschi, M. L., Jones, D. S., Panzeri, H., Gremião, M. P. D., De Freitas, O., & Lara, H. G. (2007). Semisolid systems containing propolis for the treatment of periodontal disease: In vitro release kinetics, syringeability, rheological, textural, and mucoadhesive properties. *Journal of Pharmaceutical Sciences*, *96*(8), 2074–2089.

Chugh, I., Seth, N., Rana, A. C., & Gupta, S. (2012). Oral sustained release drug delivery system: An overview. *International Research Journal of Pharmacy*, *3*, 57–61.

Collett, J., & Moreton, C. (2001). Modified-release peroral dosage form. In M. E. Aulton (Ed.), *Pharmaceutics: The science of dosage form design* (2nd ed., pp. 289–306). London: Churchill Livingstone.

Dimasi, J. A., Hansen, R. W., & Grabowski, H. G. (2003). The price of innovation: New estimates of drug development costs. *Journal of Health Economics*, *22*, 151–185.

FDA—Food and Drug Administration. (1938). *Food, drug, and cosmetic act—Pub L n° 75–717, 52 Stat. 1040*. Silver Spring: U.S. Department of Health and Human Services.

FDA—Food and Drug Administration. (2004). *Innovation or stagnation: Challenge and opportunity on the critical path to new medical products*. Silver Spring: U.S. Department of Health and Human Services.

Florence, A. T., & Attwood, D. (2006). *Physicochemical principles of pharmacy* (4th ed.). London: Pharmaceutical Press.

Gennaro, A. R. (2000). *Remington: The science and practice of pharmacy* (20th ed.). Philadelphia: Lippincott Williams & Wilkins.

Maurin, M. B., Hussain, A. A., & Dittert, L. W. (2006). Dosage form design: A physicochemical approach. In J. Swarbrick (Ed.), *Enciclopedy of pharmaceutical technology* (1st ed., pp. 939–947). New York: Informa Healthcare.

Medlicott, N. J., Rathbone, M. J., Tucker, I. G., & Holborow, D. W. (1994). Delivery systems for the administration of drugs to the periodontal pocket. *Advanced Drug Delivery Reviews*, *13*, 181–203.

Mignani, S., El Kazzouli, S., Bousmina, M., & Majoral, J. P. (2013). Expand classical drug administration ways by emerging routes using dendrimer drug delivery systems: a concise overview. *Advanced Drug Delivery Reviews*, *65*(10), 1316–1330.

Panchagnula, R., & Thomas, N. S. (2000). Biopharmaceutics and pharmacokinetics in drug research. *International Journal of Pharmaceutics*, *201*, 131–150.

Pena, S. D. J. (2002). *Homo brasilis: Genetic, linguistic, historical and socioanthropologic aspects of the formation of the Brazilian people*. FUNPEC: Ribeirão Preto.

Proudfoot, S., & Collett, J. (2001). Dosage regimens. In M. E. Aulton (Ed.), *Pharmaceutics: The science of dosage form design* (2nd ed., pp. 275–288). London: Churchill Livingstone.

Modification of drug release

2

2.1 Objectives

The administration of an active agent has the main objective of enabling the action for which it is intended. Many times, the conventional therapeutic regimens are characterized to use a high amount of a drug, with a high fraction excreted without exert activity. According to the definition, modified drug release means that the release of the active agent is different from traditional release. Therefore, modified release dosage forms are those whose drug release characteristics of time course and/or location are chosen to accomplish therapeutic or convenience objectives not offered by conventional forms (USP, 2014).

The term *drug delivery* covers a very broad range of techniques used to get therapeutic agents into the human (or animal) body. It refers to the technology utilized to present the drug to the desired body site for drug release and absorption, or the subsequent transport of the active ingredients across the biological membranes to the site of action.

A drug delivery system (DDS) is defined as a formulation or a device that enables the introduction of a therapeutic substance into the body and improves its efficacy and safety by controlling the rate, time, and place of release of drugs in the body. The therapeutic agent can also be an agent such as gene therapy, which will induce *in vivo* production of the active therapeutic agent. Despite the fact that gene therapy has its own special regulatory control, gene vectors may need to be introduced into the human body by novel delivery methods (Jain, 2008).

In this context, time and spatial control of drug release constitutes the main aim of controlled drug delivery from the systems and has evolved as a multidisciplinary science, with the contribution of engineering technologies and polymer science. This field is not recent, considering the publication of the Noyes–Whitney equation in 1897 (Noyes & Whitney, 1897), the investigations about drug dissolution and particle size in the 1950s, the mathematical studies about diffusion (Philibert, 2005) and drug release of Takeru Higuchi and his team in the 1960s (Higuchi, 1961, 1962, 1963) and the more precise appreciation of the underpinning technologies and biologies.

The controlled drug delivery technology is complex and has originated a terminology that is sometimes ambiguous and controversial. It has been common to use many terms to describe the sustained and controlled release dosage forms. These terms are used to identify specific products and their performance characteristics. They have obscured the capabilities of the products to the point where practitioners are often unclear about their advantages and limitations.

Therefore, it is helpful to distinguish some terms and definitions. In immediate release systems, the rate of appearance of a drug in the body is controlled by the biological absorption process; whereas in sustained release systems, this is controlled by the dosage form. In this sense, suspensions, emulsions, capsules and tablets can be sustained by

Strategies to Modify the Drug Release from Pharmaceutical Systems. http://dx.doi.org/10.1016/B978-0-08-100092-2.00002-3

DDSs. However, the purpose of the DDS is to decrease the dosing frequency by at least a factor of 2, and in most cases an attempt is being made for all routes of administration to prepare at least once-daily products (Longer, Middleton, & Robinson, 1988).

In this context, to achieve a therapeutic drug concentration promptly in the body, and then to maintain that concentration for a given period of time, it is necessary that the total active agent in the formulation consist of two portions: one to provide the initial loading dose (burst) and another to provide the maintenance or sustained dose. The initial dose provides a rapid onset of the desired therapeutic response, while the maintenance portion remaining in the dosage form is released at a slow but defined rate (according to zero-order kinetics). It is very important to consider that the kinetics of absorption of the maintenance dose will be characterized by the same zero-order release rate constant.

Moreover, the rate at which the maintenance dose is released from the dosage form, and hence the input of the drug into the body, must be equal to the rate of drug output from the body when the concentration of the drug in the body is at the required therapeutic value.

Therefore, considering rate and time, the difficulty of developing an ideal modified or controlled DDS is evident. The physiological conditions associated with the site of administration and the elimination rate constant of a given drug vary from patient to patient and conduce to imprecise controlled rates. Therefore, many modified-release products are not ideal controlled-release dosage forms. However, such products provide modified release in which the initial priming dose of the active agent is released immediately and the maintenance dose is then released slowly, thereby resulting in a therapeutic drug/tissue concentration that is prolonged but not kept constant (Collett & Moreton, 2001).

With the increased interest in sustained release drug delivery originated the concept of controlled release systems, carrying a superior performance concept where predictive control over the release pattern and subsequent tissue or blood levels can be achieved. Based on a simple one-compartment model, the system should be equivalent to an intravenous infusion of the drug (zero-order input). This analysis assumes that pharmacokinetics mimics pharmacodynamics; this assumption is invalid for a great number of drugs, particularly peptides and proteins. Moreover, it assumes that the biological noise is small enough that zero-order delivery can be distinguished from other kinetic input functions. The optimal rate of the drug input may not be zero order for all active agents and disease states. For example, it is known that circadian rhythm influences enzyme induction and inhibition, and that down regulation of receptors may modify drug response or metabolism (Longer et al., 1988).

Therefore, the tendency is for the authors to try to define, each in their own way, just how the release occurs or should occur. Thus, there is a great variety of terms used to describe these systems (Collett & Moreton, 2001):

(1) Delayed release (This indicates that the drug is not being released immediately following administration, but at a later time (e.g., enteric coated tablets, pulsatile-release capsules).)

(2) Repeat action (This indicates that an individual dose is released fairly soon after administration, and second or third doses are subsequently released at intermittent intervals.)

(3) Prolonged action/release (This indicates that the drug is provided for absorption over a longer period of time than for a conventional dosage form. However, there is an implication that onset is delayed because of an overall slower release rate from the dosage form.)
(4) Sustained action/release (This indicates an initial release of drug sufficient to provide a therapeutic dose soon after administration, and then a gradual release over an extended period.)
(5) Extended release (The dosage forms release the drug slowly, so that plasma concentrations are maintained at a therapeutic level for a prolonged period of time, usually between 8 and 12h.)
(6) Controlled release (The dosage forms release the drug at a constant rate and provide plasma concentrations that remain invariant with time.)

Other terms are also utilized and include "programmed release," "slow release," "timed release," "double release," "gradual release," "delayed action," "long acting," and "retard."

The term "controlled release" is general and encompasses all the others without specifying the type of control that occurs and how it is reached. Actually, this term has more fidelity, considering that the controlled DDS needs to provide the release of an active agent with control according to the objective and independent of environmental conditions.

Thus, the development of highly controllable DDS that allow the release of an appropriate amount of a drug at a specific site of action becomes an important field of investigation. The number of researches for developing methods of releasing the drug continuously for prolonged and controlled time periods and in specific sites has increased (Wise, 2000).

Controlled DDSs are being used with the aim of accurately delivering small amounts of drugs in small volumes to specific locations in the body, optimizing the drug utilization. The selection and development of an efficacious system is dependent on physicochemical and biopharmaceutical properties of the active agent and the desired concentration profile in the plasma or site of action (Maurin, Hussain, & Dittert, 2006). The aim is to reduce local or systemic side effects and to cure or control the condition in the shortest time possible, employing the smallest quantity of active agent.

The biological barriers of organisms are a challenge to conduce the success of a DDS. The identification and understanding of the mechanisms of these barriers are limiting factors for controlled drug delivery, and the use of devices/systems can circumvent the environmental/biological constraints. Moreover, the success of a DDS is dependent on its ability to achieve this and minimize bioincompatibilities or interactions with the site of administration.

In this context, the absorption and metabolic barriers that may be present in the site of administration should be considered. For example, other routes of administration can provide systematic availability when oral administration is contraindicated due to a physiologic condition. The buccal, transdermal, rectal, ophthalmic, nasal, and pulmonary routes provide advantages, allowing a concentration-time profile that approaches intravenous dosing profiles (Maurin et al., 2006).

Therefore, alterations in expected rates of absorption, distribution, metabolism, or excretion may be results of problems on all routes of administration and, subsequently, on drug-blood levels, due to the pathological disease states and drug-induced pathologies. In this context, the main biological factors associated with various routes of administration that affect the performance of DDS are (Longer et al., 1988)

(1) Buccal and sublingual (salivary dissolution and dilution, regional variations in keratinization and blood flow, generally low permeability than other oral routes)
(2) Stomach (pH)
(3) Small intestine (motility)
(4) Colon (hepatic first-pass metabolism; luminal, membrane, and intracellular enzymes; active transport mechanisms and presence of bacteria)
(5) Transdermal (low permeability, potential allergic responses, pathological states, potential binding in epidermis)
(6) Parenteral (biocompatibility, involvement of reticuloendothelial system, systemic inactivation or metabolism, vascular patency and potential lymphatic uptake)

These obstacles are considered by drug delivery technology to bring both therapeutic and commercial value to health care products. The market share of pharmaceutical companies has been loosening to generic competitors. The need for development of new dosage forms containing "old" drugs, and utilizing novel forms of delivery, is also a reality. The improving of dosage forms was very important and nowadays, we have drug delivery devices that are truly therapeutic systems.

2.2 Advantages and disadvantages

During the twentieth century, pharmaceutics was concerned with the science and practice of the manufacture of dosage forms at small and large scale and with the preparation of galenicals (Florence, 2011). The development in tabletting and the advances in material sciences have provided oral controlled delivery systems and a large range of materials to develop systems that are micro- or nanostructured. In addition, the nanotechnology has aided progress towards defining and refining the behavior of the DDS *in vivo*. Therefore, the concept of a therapeutic system was introduced as *a drug-containing preparation or dosage form which releases one or more drugs continuously in a predetermined pattern for a fixed period of time, either systemically or to a specified target organ* (Heilmann, 1978).

These strategies have the objective of regulating the rate of drug release, sustaining the duration of therapeutic action and/or targeting the delivery of a drug to a specific tissue. As result, DDS can provide one or more of the following benefits (Chien, 1992):

- Controlled administration of a therapeutic dose at a desirable rate of release
- Maintenance of drug concentration within an optimal therapeutic range for prolonged duration of treatment
- Maximization of efficacy–dose relationship
- Less accumulation of active agent in the organism
- Reduction of adverse side effects
- Protection of active agent against eventual degradation in the biological fluids
- Economy of active agent
- Minimization of the need for frequent dose intake

- Continuous treatment, without nocturnal administration
- Enhancement of patient compliance

Controlled drug delivery has aroused great interest. Many specialized publications and an outstanding number of research have generated numerous strategies that seek the exact control of release of an active agent at the target site. Mechanic, physicochemical and chemical methods have been developed to control both the access and the bioavailability of an active agent in the place of action.
On the other hand, controlled DDSs display some disadvantages, such as

(1) Difficulty (impossibility) of rapid discontinuation of pharmacological action in the event of poisoning or intolerance
(2) Risk of drug accumulation in the case of slow elimination rate
(3) Reproducibility of action dependent on gastric emptying
(4) Weak or null efficacy in the case of difficulty of absorption
(5) Difficulty of adaptation of dosage to different interindividual pharmacokinetics
(6) Release kinetic dependent on the integrity of the dosage form
(7) Size of the preparation (sometimes very big)
(8) Sometimes weak bioavailability
(9) Higher costs compared to conventional dosage forms

An ideal controlled DDS should not be highly sensitive to environmental factors, should be predictable by physicochemical principles, may be used with various active agents and doses, possesses the ability to increase or maintain the physical and chemical stability of the active agent, and finally, should be an optimized system of active agents, ensuring maximum efficiency, security, and reliability.

2.3 History

The goal of therapeutic systems is to treat a disease in a consistent and predictable fashion. A correlation between the optimal concentration of an active agent at the target site and the resulting pharmacological effect for the appropriate time of duration is assumed. The modified and/or controlled drug release produces immediate benefits, making the treatment more effective. Moreover, DDS have also received a lot of attention because they can be used to make old drugs more effective and useful. The controlled release technology has been used for life cycle management of various drugs whose patent protection had expired, as well as new types of drugs, biologics and genes (Yun, Lee, & Park, 2014).

Many efforts have been performed to achieve the accurate modulation of delivery of most active agents. Over seven decades, drug delivery technology has demonstrated significant advances. Many companies have helped accelerate the efforts, due to the fact that the fraction of medicines containing new drugs has steadily decreased, reflecting the tremendous cost required to bring new chemical entities to the marketplace.

The development of controlled release drug delivery devices is extremely complex and interwoven, and it requires a team approach. The limitations of the traditional dosage forms previously described and a basic pharmacokinetic understanding of a given drug's disposition in the human (or animal) body have been essential for the development of practically all controlled release delivery devices (Ritschel, 1988). Therefore, DDS have advanced for more than 60 years.

In the early of 1900s, the age of botanicals was losing space for a more modern approach based on developments of synthetic chemistry. Following World War II, medical technology rapidly improved as a consequence, as much of the military technology that had been developed was redirected towards medicine. Devices more complicated and more critical to patient care were the result of technological advances in electronics, materials and engineering design (Stigi, 1988).

Nowadays, personalized medicines are the aim of the drug delivery scientist, looking for more biologically precise and accurate controlled delivery systems. However, the history shows that the strategies to control the drug release from therapeutic systems came from the macro, micro and nano polymeric eras to the biological DDSs era.

The first attempts to modify the release of an active agent were observed with the development of coated pills to mask the undesirable flavor of drugs. Metallic materials were used, and retardation of the drug release was observed. Further, between 1940 and 1950, the first modified DDSs were created. They were *peroral* dosage forms with properties to release a dose of active agent in the stomach and another dose in the intestine, or they avoided the gastric release. As previously discussed, these systems were sensible to physiological variables (e.g., differentiated stomach emptying). Therefore, after 1950, the history of controlled release technology is divided into three time periods:

- 1st period from 1950 to 1970
- 2nd period from 1970 to 1990
- 3rd period from 1990 onward

The 1st period was named the period of sustained/prolonged drug release, in which wax and hydrophobic polymers were the main materials used. The lack of information about the physiological and anatomic barriers of organisms constituted an important difficulty for developing new systems.

The year 1952 marked the introduction by Smith Kline & French (SKF) of the Spansule® system, one of the first dosage forms for prolonged action of a therapeutic agent. The Spansule® was the result of SKF evaluations about the therapeutic advantages of coatings for tablets. Searches were also started for chemical or mechanical techniques that would gradually release drugs for a predictable therapeutic effect over an entire day or night.

By 1949, the possibility to fill capsules with pellets coated with medication that would dissolve at different times was realized. Actually, microcapsules containing drug pellets surrounded by coating of variable thickness were developed to prolong release. The implementation and adaptation to large scale of manufacturing was complex, and by 1952, the time-released capsule was used and marketed. Dexedrine (dextroamphetamine sulfate), a drug for treating psychiatric patients of certain types, patients suffering from depression, fatigue or listlessness, and patients afflicted with narcolepsy, was the first active agent incorporated to the Spansule® system.

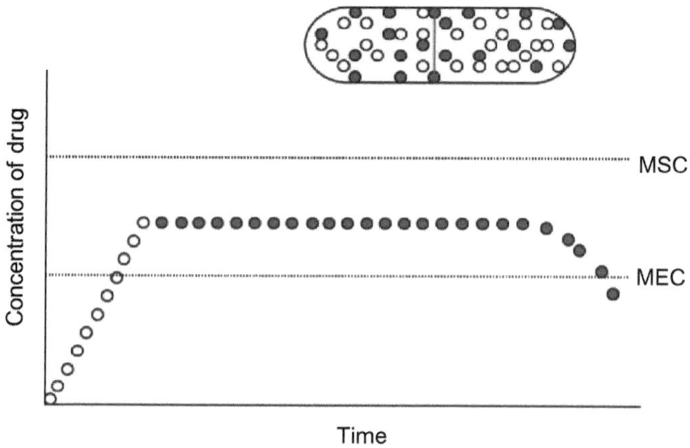

Figure 2.1 Graphical scheme of drug delivery profile from the Spansule® system (GlaxoSmithKline company). MSC, maximum safety concentration; MEC, minimum effective concentration.

More than 35,000h of work were required to develop the Spansule®. It provided a novel drug delivery form, which quickly released the required initial dose and then slowly and gradually released many extremely small doses to maintain a therapeutic level lasting from 10 to 12h (Figure 2.1), providing all-day or all-night therapy with just one administration/dose. These discoveries led to important advances in the sustained drug delivery technology.

Subsequently, SKF (nowadays GlaxoSmithKline) marketed a number of other sustained-release preparations, one of which is the cold remedy named Contact®. It was launched in 1960 and later became the world's leading cold medicine.

Coatings that provided site-control of drug delivery, such as enteric coatings, were also developed in Europe by Rohm & Haas in the late 1950s. Furthermore, the use of silicone rubber as a carrier for prolonged drug delivery was proposed in the mid-1960s by Judah Folkman and David M. Long (Folkman & Long, 1964). This work proposed the original concept of an implanted (zero-order) reservoir drug delivery device. The authors were circulating rabbit blood inside a Silastic® (silicone rubber) arteriovenous shunt and discovered that if they exposed the tubing to anesthetic gases on the outside, the rabbits would fall asleep (Folkman, Long, & Rosenbau, 1966). They proposed that short, sealed segments of such tubing containing a drug could be implanted, and if the silicone didn't change in dimensions or composition, the implant would become a constant-rate drug delivery device.

Meanwhile, Alejandro Zaffaroni, a synthetic drug chemist and entrepreneur of Palo Alto (CA, USA) founded a company in the late 1960s focused on the concept of controlled drug delivery. He was stimulated by Folkman's work as well as by his own vision, and thinking about the concept of zero-order delivery and controlled delivery devices, founded Alza (first two letters of his first and last names). Alza Laboratories was a pioneer and accelerated the development of drug delivery devices due to the substantial decline in the development of new drug entities.

In the 1960s, poly(glycolic acid) (PGA) was synthesized and patented at Davis & Geck (Cyanamid Company) for use as a degradable suture (Schmitt & Polistina, 1967). Further, poly(lactic acid) acid (PLA) was studied and lactic acid was added to the composition of PGA to make the degradable poly(lactic-co-glycolic acid), PLGA, to be used as suture material (Vicryl® by Ethicon). The polymerization of glycolic acid was made practical by Kureha Company (Japan), which developed the glycolide cyclic ester, conducing to the efficient production of PLA and PLGA (Hoffman, 2008). In the last years of the 1960s, peptide drugs were added to PLA by Du Pont Chemical Company researchers George Boswell and Richard Scribner. Microparticle and pellet depot DDSs were fabricated from them (Boswell & Scribner, 1973).

A number of advanced methods were developed for controlled drug delivery during this 1st period. However, these techniques were not fully implemented until the beginning of the 1970s.

The 2nd period was involved in the determination of the needs of controlled drug delivery, and the concept of therapeutic systems was created. This was possible due to the understanding of barriers for many routes of administration.

Biotechnology and molecular biology began to play a significant role in the drug discovery process, increasing the understanding of the etiology of many diseases. The modified and controlled delivery of potent molecules (e.g., steroids, peptides, and proteins) was studied, and the technology evolved from the macroscopic, zero-order devices in the 1970s and 1980s, to "sustained release" macro/microscopic biodegradable depot systems that first reached the clinic in the mid- to late 1980s. Furthermore, the advances in computer screening, predictive software, combinatorial chemistry and high throughput screening contributed to accelerating the identification of new therapeutic compounds.

During this period, many important events contributed to the evolution of DDSs. Pharmetrics was founded as a new company of design and development of the different zero-order DDSs for Alza. In 1970, Alan Michaels, a former chemical engineering professor from MIT and the founder and president of Amicon Corporation, was invited to be president of Pharmetrics. In 2006, Alza become part of Johnson & Johnson (Hoffman, 2008).

This period was characterized by researching the control of delivery of active agents, and the systems reached different levels of delivery control quality. The first controlled drug delivery devices developed by Alza were macroscopic in scale. For example, Ocusert®, an ophthalmic insert, was developed to release pilocarpine (anti-glaucoma drug) at a constant rate in the eye. Progestesert®, an intrauterine device (IUD), was developed to release progesterone (contraceptive steroid) at a constant rate in the uterine cavity. They used poly(ethylene-co-vinyl acetate), polyEVA, as the rate-controlling membrane in a constant-rate, reservoir drug delivery device, producing a zero order. Moreover, a contraceptive subcutaneous implant comprised of six silicone rubber tubes (cross-linked polydimethylsiloxane, PDMS) filled with a contraceptive steroid, levo-Norgestrel, called Norplant®, was developed. A similar device, Implanon®, using polyEVA as the rate-controlling membrane, was brought by Organon and approved for clinical use by the US Food and Drug Administration (FDA) in 2006 (Hoffman, 2008).

In 1971, Alza issued one of the first patents on the controlled delivery skin patch, which was called a "Bandage for Administering Drugs" (Zaffaroni & Calif, 1971). This system was a membrane-controlled reservoir device that exhibited zero-order release kinetics. PolyEVA or porous polypropylene (PP) (the pores could be filled with an oil or wax) was the most commonly used rate-controlling membrane (RCM). The RCM resistance to drug diffusion was designed to be greater than that of the *stratum corneum*. The first skin patch product contained scopolamine, an anti-motion sickness drug. Alza was interested in this anti-motion sickness drug patch for the new space program of President Kennedy, and could get government funding to develop the patches for use in space travel (Hoffman, 2008).

In the mid-1970s, hydrophobic polymer matrices and PLGA microparticles were developed to deliver a "fragile" biomolecular drug, such as a protein (e.g., steroids and peptides). As a consequence, in the late 1970s to 1980s, polymer–drug conjugates ("nanotherapeutics") were introduced. A contraceptive doughnut-shaped device for vaginal insertion was developed by the World Health Organization (WHO). It was made of an internal doughnut-shaped core of a contraceptive steroid blended with silicone rubber, and coated with polymer as the RCM (Hoffman, 2008).

During the early 1980s, the first clinically approved, injectable, degradable, microparticle depot DDS (Decapeptyl® LP) was developed by DebioPharm. This product is a [D-Trp-6 LHRH]/PLGA microparticle for the treatment of prostate cancer and was launched in the clinic in Europe in 1986. Three years later, Syntex and Takeda-Abbott Pharmaceutical Company marketed this product as the Lupron® depot (Hoffman, 2008; Kent, Lewis, Sanders, & Tice, 1987).

Injectable, degradable, depot DDSs of PLA or PLGA were also developed in the late 1980s. These depot systems are phase-separated, macroscopic "masses" formed by subcutaneous or intramuscular injections of drug/polymer/solvent systems. Thermally responsive aqueous dispersions of di-block and tri-block copolymers of PLGA-PEG (Re-Gel®, MacroMed Inc.) were developed during the same period. Moreover, a macroscopic depot DDS based on a degradable polyorthoester (POE) was developed by Alza and called Chronomer® (Hoffman, 2008).

Osmotic pump capsules were also developed by Alza at the beginning of the 1980s and into the 1990s. The osmotic systems (e.g., Oros®) were developed for controlled release of drugs in the gastrointestinal tract, with one of their first patents issued in 1974 (Theeuwes, Calif, Higuchi, & Kans, 1974). In addition, during the 1990s, implanted osmotic delivery systems (e.g., Duros®) were also developed. Both the osmotic capsules were diffusion-controlled devices with a RCM (cellulose acetate, sometimes blended with a small amount of a low molecular weight of polyethylene glycol) under constant rate of water diffusion into the capsule.

The technology of biodegradable polymers of poly(hydroxy acids) developed for sutures in the 1960s and 1970s was also adopted by drug delivery researchers in the 1970s and reached the clinic in the 1980s (Hoffman, 2008).

The development of liposomes and micelles containing poly(ethylene glycol) (PEG) was a very important marker between the 1980s and 1990s, and started the 3rd period, which is considered the modern age of controlled release. The development of biocompatible polymers and polymeric carriers led to numerous products in the market. Some examples are described below.

The Geomatrix® oral DDS of Skye Pharma, a swelling/gelling hydrogel system, was developed by Colombo (Peppas & Colombo, 1997) at the University of Padua in Italy. The Prolease® process for fabricating uniform-sized, drug-loaded PLGA microparticles was developed by Enzytech (a Langer spin-off company) (Gombotz, Healy, & Brown, 1991). The novelty was the use of an ultrasonic sprayer and a liquid nitrogen and frozen ethanol bath to freeze the uniform-sized particles and remove the solvent. Previously, emulsification and solvent evaporation or phase separation techniques had constituted the main techniques used to prepare degradable, drug-loaded microparticles.

Biodegradable polyester based on a copolymer of poly(ethylene glycol terephthalate) (PEG-T) and poly(butylene terephthalate) (PBT) was first synthesized and studied as microparticle depot formulations for delivery of α-Interferon (Locteron®, Octoplus Company) (Hoffman, 2008).

As previously described, the concept of nanotherapeutics (polymer–drug conjugates) independently arose at various places around the world in the mid- to late 1970s, and took center stage after 1990. Three key technologies are considered the major factors that stimulated the immense activity and clinical success of nanotherapeutics from the past to the present (Hoffman, 2008; Bader & Putnam, 2014; Yun et al., 2014):

(1) The concept of PEG-conjugated drugs or drug carriers (PEGylation)
(2) Conjugation of cell membrane receptor antibodies, peptides or small molecule cell ligands to the polymer carrier (active targeting)
(3) The discovery of nanoscale carriers that are entrapped within solid tumors due to leaky vasculature of the fast-growing tumor (enhanced permeation and retention effect)

The concept of PEGylation started to be constructed in the late 1960s, when it was observed that the recombinant protein drugs that were just being developed showed enhancement of both the circulation time and the stability (against enzyme attack or immunogenic recognition) by linking PEG (Davis, 2002). The discovery led to the foundation of Enzon, a PEGylation company, at the beginning of the 1980s. This was the start to the clinical studies with PEGylated enzymes (e.g., asparaginase and glutaminase, which metabolized asparagine and glutamine, essential nutrients for leukemic cancer cells). The other important PEGylation company founded was Shearwater Polymers (Huntsville, AL, USA), collaborating to introduce a number of PEGylated recombinant protein products to the clinic (Hoffman, 2008).

Around the world, researchers had the idea of a targeted, polymer–drug conjugate (Ringsdorf, 1975; Duncan, Cable, Lloyd, Rejmanova, & Kopecek, 1983; Rejmanova, Kopecek, Pohl, Baudys, & Kostka, 1983; Cuchelkar & Kopecek, 2006). A new polymer carrier called poly(hydroxypropyl methacrylamide) (PHPMA) was synthesized and found to be non-immunogenic and able to deliver drugs such as doxorubicin and other small anticancer molecules. The drug–polymer conjugates could be actively targeted with ligands such as galactose, an asialo-glycoprotein membrane receptor ligand for hepatocytes, for liver cancer treatment. In 1984, the "Enhanced Permeation and Retention" effect (Iwai, Maeda, & Konno, 1984) was discovered. The researchers were carrying out animal studies with styrene-maleic anhydride conjugated to the anticancer peptide drug neocarzinostatin, and they had labeled the conjugate with a dye. They observed the accumulation of the dye within the tumor tissue, and they concluded that the rapidly forming vasculature in such solid tumors was "leaky," while the lymph drainage system was not yet working efficiently, which led to its entrapment or accumulation within the tumor tissue. Today, this effect was observed to be only effective close to the leaky vessels, and not throughout the tumor, due perhaps to the low diffusion coefficient of the nanocarriers within the tumor's extravascular tissues (Hoffman, 2008). Nowadays, other nanoscale DDSs have been developed, including PEGylated polymeric micelles and liposomes.

In the early 1990s, the emergence of important nucleic acid drugs, such as plasmid DNA (pDNA) and antisense ODNs ("oligos"), led to the development of cationically condensed pDNA or ODN nanoparticles. Both cationic polymers and cationic liposomes were used to condense the nucleic acid drugs. The resultant complexes are called polyplexes or lipoplexes, respectively, and a number of companies (e.g., Alnylam, Roche, Merck, Calando) have been involved in clinical trials for delivery of siRNA from these systems (Hoffman, 2008).

Nowadays, nanoscale albumin-based drug carriers, liposomal formulations, dendrimers, dendronized polymers and other hyper-branched polymers also have been studied. Drug-eluting coronary stents and delivery systems that stimulate physiological patterns have been approved. A drug-eluting stent is a clinical application of drug-loaded surface coatings (Westedt et al., 2006). Cypher® (Johnson & Johnson) and Taxus® (Boston Scientific) are stents designed to release a smooth muscle cell (SMC) antagonist (rapamycin or paclitaxel, respectively) that retards or prevents SMC proliferation, the main cause of vessel stenosis. Drug particles are blended with the polymer, which in the Cypher® stent is a polyblend of poly(butyl methacrylate) and polyEVA, while that used in the Taxus® stent is a novel tri-block copolymer of polystyrenepolyisobutylene-polystyrene (Hoffman, 2008).

Some polymers have been developed and clinically applied as mucoadhesive DDS with the property of site control release, increasing retention times at the surfaces of mucosal tissues such as eye, nose and vagina, and within the gastrointestinal tract. Acrylic acid polymers, chitosan and cellulose derivatives are some examples (Mathiowitz, Chickering, & Lehr, 1999). Moreover, thermoresponsive polymers, such as poloxamers, have been used to improve the properties of these systems (Bruschi et al., 2007, 2008; Jones, Bruschi, De Freitas, Gremião, & Lara, 2009; Pereira, Godoy, Svidzinski, & Bruschi, 2013).

Another very important evolution of drug delivery was dependent on the growing understanding about DNA sequences encoded for each disease. This has provided the possibility of predicting the precise therapeutic regimen for optimum treatment of that disease in each patient. The increased knowledge of cell biology and DNA has enabled the development of nanoscale modified DDSs that are serum stable and efficiently taken up by specific cells, and that can then escape the endosome and target specific sites and pathways within the cells.

It is clear that the progress in strategies to control the drug release from pharmaceutical systems made during the last 60 years has been significant. Analyzing the results, it is possible to organize the evolution into two main groups. The first group is composed of knowledge obtained up to and during the 1970s, in which are the basics of controlled release (Yun et al., 2014):

- Physicochemical properties of dissolution, diffusion, osmosis, and ion-exchange influencing the drug delivery mechanism
- Oral delivery with reduced number of administration of doses (once a day, twice a day)
- Transdermal delivery with once-a-day or once-a-week administration

The second group is constituted by the knowledge obtained from the 1980s until the 2000s, in which the concept of smart delivery systems originated. The main points highlighted are the order of release, hydrogels and smart polymers, peptide and protein delivery, nanostructured systems and materials of natural origin (Yun et al., 2014). Nowadays, the aim is to modulate the systems, considering these last advances, to reach the next generation of DDSs.

The past and the present must both be considered to predict the future of drug delivery. The design of new and potent biomolecular drugs with minimal side effects, together with the increased ability to control the efficiency and specificity of the drug delivery process, have conducted the field of DDSs to become more biological and less materials-oriented in character.

References

Bader, R. A., & Putnam, D. A. (2014). *Engineering polymer systems for improved drug delivery*. New York: Wiley.

Boswell, G., & Scribner, R. (1973). Polylactide-drug mixtures, US Patent 3,773,919.

Bruschi, M. L., Jones, D., Panzeri, H., Gremião, M. P. D., De Freitas, O., Jones, D. S., et al. (2008). Precursor system of liquid crystalline phase containing propolis microparticles for the treatment of periodontal disease: Development and characterization. *Drug Development and Industrial Pharmacy*, *34*, 267–278.

Bruschi, M. L., Jones, D., Panzeri, H., Gremião, M. P. D., De Freitas, O., & Lara, E. H. G. (2007). Semisolid systems containing propolis for the treatment of periodontal disease: in vitro release kinetics, syringeability, rheological, textural, and mucoadhesive properties. *Journal of Pharmaceutical Sciences*, *96*(8), 2074–2089.

Chien, Y. W. (1992). *Novel drug delivery systems*. New York: Marcel Dekker.

Collett, J., & Moreton, C. (2001). Modified-release peroral dosage form. In M. E. Aulton (Ed.), *Pharmaceutics: The science of dosage form design*. (2nd ed., pp. 289–306). London: Churchill Livingstone.

Cuchelkar, V., & Kopecek, J. (2006). Polymer-drug conjugates. In I. Uchegbu, & A. Schatzlein (Eds.), *Polymers in drug delivery* (pp. 155–182). London: Taylor & Francis.

Davis, F. F. (2002). The origin of PEGnology. *Advanced Drug Delivery Reviews*, *54*(4), 457–458.

Duncan, R., Cable, H. C., Lloyd, J. B., Rejmanova, P., & Kopecek, J. (1983). Polymers containing enzymatically degradable bonds. 7. Design of oligopeptide side-chains in poly[n-(2-hydroxypropyl)methacrylamide] co-polymers to promote efficient degradation by lysosomal-enzymes. *Macromolecular Chemistry and Physics*, *184*, 1997–2008.

Florence, A. T. (2011). A short history of controlled drug release and an introduction. In C. G. Wilson, & P. J. Crowley (Eds.), *Controlled release in oral drug delivery* (pp. 1–26). London: Springer.

Folkman, J., & Long, D. M. (1964). The use of silicone rubber as a carrier for prolonged drug therapy. *Journal of Surgical Research*, *4*, 139–142.

Folkman, J., Long, D. M., & Rosenbau, R. (1966). Silicone rubber—A new diffusion property useful for general anesthesia. *Science*, *154*, 148–149.

Gombotz, W., Healy, M., & Brown, L. (1991). Very low temperature casting of controlled release microspheres, US Patent 5,019,400.

Heilmann, K. (1978). *Therapeutic systems—Pattern-specific drug delivery: concept and development*. Stuttgart: Thieme.

Higuchi, T. (1961). Rate of release of medicaments from ointments bases containing drugs in suspension. *Journal of Pharmaceutical Sciences*, *50*(10), 874–875.

Higuchi, W. I. (1962). Analysis of data on the medicament release from ointments. *Journal of Pharmaceutical Sciences*, *51*(8), 802–804.

Higuchi, T. (1963). Mechanism of sustained-action medication. *Journal of Pharmaceutical Sciences*, *52*(12), 1145–1149.

Hoffman, A. S. (2008). The origins and evolution of "controlled" drug delivery systems. *Journal of Controlled Release*, *132*, 153–163.

Iwai, K., Maeda, H., & Konno, T. (1984). Use of oily contrast medium for selective drug targeting to tumor: Enhanced therapeutic effect and X-ray image. *Cancer Research*, *44*(5), 2115–2121.

Jain, K. K. (2008). Drug delivery systems – An overview. In K. K. Jain (Ed.), *Drug delivery systems* (pp. 1–50). Totowa: Humana Press.

Jones, D. S., Bruschi, M. L., De Freitas, O., Gremião, M. P. D., & Lara, E. H. G. (2009). Rheological, mechanical and mucoadhesive properties of thermoresponsive, bioadhesive binary mixtures composed of poloxamer 407 and carbopol 934P designed as platforms for implantable drug delivery systems for use in the oral cavity. *International Journal of Pharmaceutics*, *372*, 49–58.

Kent, J., Lewis, D., Sanders, L., & Tice, T. (1987). Microencapsulation of water soluble active polypeptides, US Patent 4,675,189.

Longer, M. A., Middleton, D. L., & Robinson, J. R. (1988). Conventional routes of delivery and the need for devices. In P. Tyke (Ed.), *Drug delivery devices: Fundamentals and applications* (pp. 3–16). New York: Marcel Dekker.

Mathiowitz, E., Chickering, D., & Lehr, C.-M. (1999). *Bioadhesive drug delivery systems—Fundamentals, novel approaches, and development*. New York: Marcel Dekker.

Maurin, M. B., Hussain, A. A., & Dittert, L. W. (2006). Dosage form design: a physicochemical approach. In J. Swarbrick (Ed.), *Enciclopedia of pharmaceutical technology* (1st), (pp. 939–947). New York: Informa Healthcare.

Noyes, A. A., & Whitney, W. R. (1897). The rate of solution of solid substances in their own solutions. *Journal of the American Chemical Society*, *19*, 930–934.

Peppas, N. A., & Colombo, P. (1997). Analysis of drug release behavior from swellable polymer carriers using the dimensionality index. *Journal of Controlled Release*, *5*, 35–40.

Pereira, R. R. A., Godoy, J. S. R., Svidzinski, T. I. E., & Bruschi, M. L. (2013). Preparation and characterization of mucoadhesive thermoresponsive systems containing propolis for the treatment of vulvovaginal candidiasis. *Journal of Pharmaceutical Sciences*, *102*(4), 1222–1234.

Philibert, J. (2005). One and a half century of diffusion: Fick, Einstein, before and beyond. *Diffusion Fundamentals*, *2*, 1.1–1.10.

Rejmanova, P., Kopecek, J., Pohl, J., Baudys, M., & Kostka, V. (1983). Polymers containing enzymatically degradable bonds. Degradation of oligopeptide sequences in N-(1992-hydroxypropyl)methacrylamide co-polymers by bovine spleen cathepsin-B. *Macromolecular Chemistry and Physics*, *184*, 2009–2020.

Ringsdorf, H. (1975). Structure and properties of pharmacologically active polymers. *Journal of Polymer Science*, *51*, 135–153.

Ritschel, W. A. (1988). Pharmacocinetic and biopharmaceutical aspects in drug delivery. In P. Tyle (Ed.), *Drug delivery devices: Fundamentals and applications* (pp. 17–80). New York: Marcel Dekker.

Schmitt, E., & Polistina, R. (1967). Surgical sutures, US Patent 3,297,033.

Stigi, J. F. (1988). The regulation of drug delivery devices. In P. Tyle (Ed.), *Drug delivery devices: Fundamentals and applications* (pp. 81–133). New York: Marcel Dekker.

Theeuwes, F., Calif, L.A., Higuchi, T., Kans, L. (1974). Osmotic dispensing device for releasing beneficial agent. US Patent 3,845,770.

USP. (2014). *United States Pharmacopoeia 37—National Formulary 32*. Rockville: US Pharmacopoeial Convention.

Westedt, U., Wittmar, M., Hellwig, M., Hanefeld, P., Greiner, A., Schaper, A. K., et al. (2006). Paclitaxel releasing films consisting of poly(vinyl alcohol)-graft-poly(lactide-co-glycolide) and their potential as biodegradable stent coatings. *Journal of Controlled Release*, *111*, 235–246.

Wise, D. L. (2000). *Handbook of pharmaceutical controlled release technology*. New York: CRC Press.

Yun, Y., Lee, B. K., & Park, K. (2014). Controlled drug delivery systems: the next 30 years. *Frontiers of Chemical Science and Engineering*, *8*(3), 276–279.

Zaffaroni, A., & Calif, A. (1971). Bandage for administering drugs, US Patent 3,598,122.

Classification of therapeutic systems for drug delivery

3

3.1 Introduction

Systems that display the ability for sustained drug release are studied, and several new modifications are introduced. The terms long-acting or delayed release systems sometimes overlap with controlled release systems. However, the more sophisticated control of drug release implies going beyond just the time dimension. Moreover, newer and more powerful drugs are developed continuously, and great attention is given to the methods by which these active agents are administered.

The need for prolonged and better control of drug administration is very important. The aim of disease treatment is highly dependent on the ability of the therapeutic system to selectively and effectively treat targeted cells and tissues while leaving other healthy parts of the body untouched.

New strategies of administering drugs (i.e., orally, systemically, or loco-regionally) play a critical role in the development of new therapeutic systems. In addition, the discovery of new therapeutics that can alter specific biological processes that are critical to disease pathology is very important as well. Many drug delivery systems are already available commercially, and a great part of them has the ability to control the drug release in a temporal and/or spatial way.

Therefore, after a single dose, the system can maintain the drug in the desired site of action, with a therapeutic range, improving the efficacy and safety of therapeutics (Kost & Langer, 2012). For this, necessary elements include the administration of the therapeutic system, the release of the active agent by the formulation, and the subsequent transport of the active agent across the biological membranes for systemic effects or targeted to various organs and diseases.

The therapeutic systems can introduce the active agent into the human body by various anatomical routes (gastrointestinal system, parenteral, transmucosal, transnasal, pulmonary, transdermal, and intra-osseous). The choice of the administration route depends on the effect desired, the disease and the formulation available. It is also possible to administer the agent directly to the organ affected by disease or give it systemically and targeted to the diseased organ.

The system constitutes an interface between the patient and the drug. It is very important to distinguish between the active agent and the system. In addition, the biological environment where the system containing the drug is administered constitutes another variable that needs to be considered.

The development of constant-rate drug delivery systems is a consequence of the assumption that drug concentration–effect relationships are significantly invariant as

Strategies to Modify the Drug Release from Pharmaceutical Systems. http://dx.doi.org/10.1016/B978-0-08-100092-2.00003-5

a function of time in humans. However, the body is essentially a complex network of compartments and biochemical reactions within which the desired sites of action lie.

There are clinical situations where such an approach may not be sufficient. Kost and Langer (2012) highlighted some examples: anti-arrhythmics for patients with heart rhythm disorders, gastric acid inhibitors for ulcer control, delivery of insulin for patients with diabetes mellitus, nitrates for patients with angina pectoris, selective β-blockade, birth control, general hormone replacement, immunization and cancer chemotherapy. It is very important to consider that the onset of certain diseases exhibits strong circadian temporal dependency.

Thus, drug delivery patterns from the systems need to be determined and optimized to reach the suitable amount of active agent in the site of action for the suitable time, considering the variables of the biological organism.

Therefore, many strategies and materials have been proposed. Over the last 30 years, the development of biorelated polymers for medical applications has undergone a great transition from the pragmatic approach to applying basic research principles.

Many new materials were developed for intended applications and solutions to problems related to those applications (Ottenbrite & Kim, 2002). Research groups have developed responsive systems that would more closely resemble the normal physiological process in which the amount of drug released can be affected according to physiological needs. The concept of therapeutic systems was created during the 1970s, when the control of drug release was taking place. Today, these systems have reached high levels of quality regarding the control of drug release. Importantly, the properties of an ideal drug delivery or biomedical vector are as follows (Jain, 2008, 2012):

(1) Structural control over size and shape of drug or imaging-agent cargo space
(2) Biocompatible, non-toxic polymer/pendant functionality
(3) Precise, nanoscale-container and/or scaffolding properties with high drug or imaging-agent capacity features
(4) Well-defined scaffolding and/or surface modifiable functionality for cell-specific targeting moieties
(5) Lack of immunogenicity
(6) Appropriate cellular adhesion, endocytosis and intracellular trafficking to allow therapeutic delivery or imaging in the cytoplasm or nucleus
(7) Acceptable bioelimination or biodegradation
(8) Controlled or triggerable drug release
(9) Molecular level isolation and protection of the drug against inactivation during transit to target cells
(10) Minimal non-specific cellular and blood-protein binding properties
(11) Ease of consistent, reproducible, clinical-grade synthesis

This advance conduced to the evolution and discovery of new materials. Moreover, the development during this time has made it possible to program the drug release from the systems in a passive or active way. Approaches used for achieving the suitable drug release may be physical, chemical, or biological mechanisms. Thus, the therapeutic systems for drug delivery can be classified as passive (preprogrammed), active (programmable or externally controlled), and self-programmable.

The design of controlled-release drug delivery systems takes the drug, delivery and destination into account. Moreover, the active systems also can be classified as open- or closed-loop systems. Open-loop control systems are those in which information about the controlled variable is not automatically used to adjust the system inputs to compensate for the change in the process variables. On the other hand, in closed-loop control systems, the controlled variable is detected and, as a result, the system output is adjusted accordingly. Open-loop systems are known as pulsed or externally regulated, and closed-loop systems as self-regulated.

A drug delivery system is developed to enable the introduction of a therapeutic substance into the biological organism and improves its efficacy and safety by controlling the rate, time and place of release of drugs in the body (Jain, 2008). In the second half of the 1970s, a drug delivery system was described as composed of the active(s) agent(s), the drug delivery module, the platform and the therapeutic program (Heilmann, 1978). This composition was generic and is still useful nowadays. The active agents are more powerful and optimized, using new approaches such as nanotechnology and genetic therapy.

Development of efficient therapeutic systems is of vital importance for medicine and health care, and platforms are developed and characterized with the aim of controlling drug delivery (Jones et al., 2009).

The delivery module is contained in the platform and has the function to release the active agent according to a predetermined therapeutic program. Therefore, time and/or spatial control can be possible, considering the administration route, target site and drug characteristics. New and smart materials have enabled the development of biodegradable, biocompatible, environment-responsive, targeted delivery systems (Kost & Langer, 2012; Bruschi et al., 2007, 2008; Farmer, Edgar, & Peppas, 2008; Shim & Kwon, 2012; Chan, Orme, Fricker, & Roach, 2013). The new materials and strategies have synergistically fuelled the advancement of drug delivery.

3.2 Passive, pre-regulated delivery systems

Passive systems are also considered pre-programmed or pre-regulated, in which the logical element (drug release controller) is represented by a membrane, a series of laminates or a matrix (Figure 3.1).

The rate of drug release is generally time-dependent (zero-order) or at a predictable constant declining rate (first-order) for a certain time period (Heilmann, 1978). The system is already programmed by the factory, and independent of physical, chemical, or biological *in vivo* processes.

The required concentration of the drug at the site of action over a predetermined period of time is a consequence of the equilibrium between the optimized administration, controlled drug release and elimination.

Passive therapeutic systems can control drug delivery, and the optimized (reduced) dose of drug administered to the patient reduces the problem of side effects, increasing safety, and improving the therapy and the patient compliance.

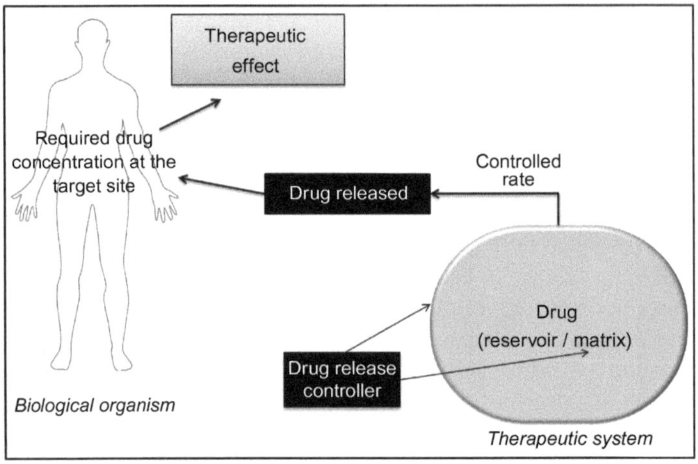

Figure 3.1 Schematic representation of pre-regulated delivery system.

3.3 Active, externally regulated delivery systems

The increasing availability of biologically derived and biocompatible synthetic materials has produced a great number of new ideas and approaches for the development of open-loop systems. The elevated degree of sophistication of many of these novel platforms has enabled the capability to control the release of active agents in response to specific endogenous or externally applied stimuli.

Programmed or externally controlled delivery systems are characterized to possess the logical element able to receive and convert a signal (e.g., electric, magnetic, thermal or ultrasonic) sent by an external source, controlling and modeling the drug release in the body (Figure 3.2).

The use of magnetic beads and a magnetic field to control drug delivery has been a reality since the studies of Kost and collaborators (Kost, Wolfrum, & Langer, 1987). They developed in vivo studies in which polymeric matrices containing insulin and magnetic beads were implanted in diabetic rats. The glucose levels were repeatedly decreased on demand by application of an oscillating magnetic field. Later, other studies showed that the insulin delivery could be controlled using the magnetic stimuli (Saslawski, Weigarten, Benoit, & Couvreur, 1988; Kost & Langer, 2012). The magnetic field characteristics and the mechanical properties of the platform are the main factors that are critical in controlling the release rates in these systems. In this sense, the extent of release enhancement increases as the magnetic-field amplitude rises (Edelman, Kost, Bobeck, & Langer, 1985).

Ultrasonically modulated systems are also feasible. They are characterized by release rates of active agents that can be repeatedly modulated from a position external to the delivery system. Both bio-erodible and non-erodible polymers were used as drug carrier matrices. Ultrasound has been evaluated as an enhancer for drug delivery through the skin and inside the organism. Controlled insulin delivery, phonophoresis (the movement of drugs through living intact skin and into soft tissue under the

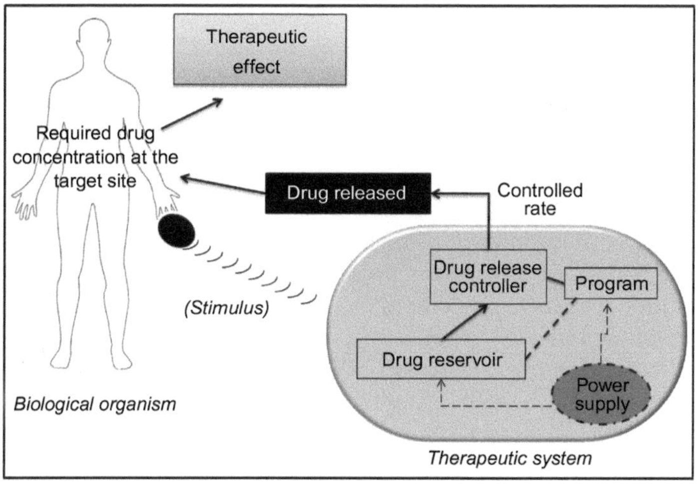

Figure 3.2 Schematic representation of open-loop therapeutic system.

influence of an ultrasonic perturbation) and the inhibition of enzymes have been studied (Kost & Langer, 2012).

The thermal stimulus has been used to control the drug delivery mainly of semi-solid systems. Thermoresponsive system hydrogels such as N-substituted polyacrylamide and those composed of poloxamer have been developed to deliver active agents of natural or synthetic origin (Bruschi et al., 2007; Jones et al., 2009; Pereira, Godoy, Svidzinski, & Bruschi, 2013; Kost & Langer, 2012).

Electrically controlled systems are characterized to modulate the transmembrane solute flux by the action of an applied electric field on the membrane and/or directly on the solute. The electrophoretic migration of a charged drug within a hydrated membrane results from the combined response to the electrical forces on the solute and its associated counterions in the adjacent electrolyte solution (Kost & Langer, 2012). Using the electric stimulus, it is possible to induce the swelling of a membrane and, therefore, to alter the effective pore size and permeability, augment the drug flux within a membrane, and induce the electrostatic partitioning of charged solutes into charged membranes. The iontophoresis uses these properties to electrically control the membrane permeability and enhanced transdermal drug delivery (Jadoul, Bouwstra, & Préat, 1999; Ciach & Moscicka-Studzinska, 2011).

Microwave irradiation and photochemical reactions are other stimuli that can also regulate drug release.

3.4 Active, self-regulated delivery systems

When the logical element is sensible and answers to the biological environment to modulate the drug release according to the momentary need, the system is classified as active, self-regulated, or self-programmable.

These systems can control the release by feedback information, without any external intervention, and use several approaches as rate-control mechanisms: pH-sensitive polymers, enzyme-substrate reactions, pH-sensitive drug solubility, competitive binding, and metal concentration-dependent hydrolysis (Kost & Langer, 2012).

They are closed-loop controlled devices (Figure 3.3) in which the release rates are adjusted by the system, in response to feedback information, without any external intervention.

The pH range of fluids in various segments of the gastrointestinal tract may provide environmental stimuli for responsive drug release. It is possible to work with materials that constitute matrices, membranes or coatings, by which the system can bypass determined environments (e.g., stomach) and release the drug in the appropriate site with a specific pH rate (e.g., intestine). Many studies have shown that in polymers containing weakly acidic or basic groups in the polymeric backbone, their charge density depends on the pH and ionic composition of the solution into which the polymer is exposed (Kost & Langer, 2012). The swelling or deswelling of the polymer is a result of the changes in the pH of the solution.

The control of insulin delivery has also been studied using closed-loop systems in which insulin and immobilized glucose oxidase can be enclosed in pH-responsive polymers (e.g., hydrogels). When the level of glucose increases, this molecule diffuses into the system and glucose oxidase catalyzes its conversion to gluconic acid, thereby lowering the pH in the micro-environment of the membrane, causing swelling and controlling the insulin release. Another way to control insulin delivery by closed-loop systems is the competitive binding process. The basic principle is the preparation of glycosylated insulins, which are complementary to the major combining site of carbohydrate-binding proteins. For example, the glycosylated insulin can bind with Concanavalin A, which is immobilized on sepharose beads. The glycosylated insulin (which is biologically active) is displaced from the Concanavalin A by glucose in

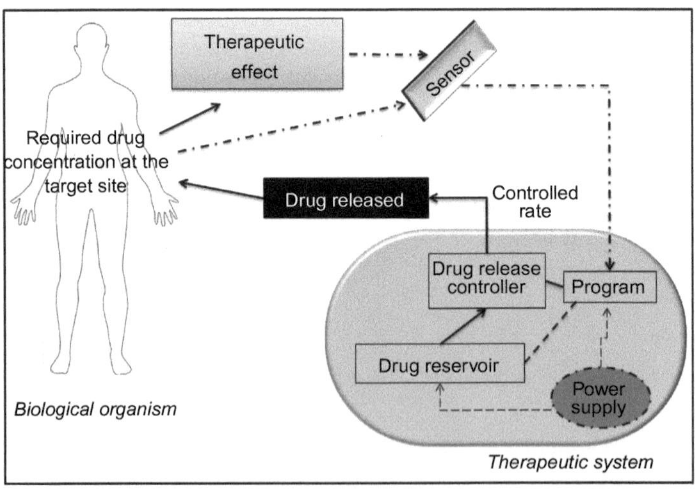

Figure 3.3 Schematic representation of closed-loop therapeutic system.

response to, and proportional to, the amount of glucose present that competes for the same binding sites. Moreover, by encapsulating the glycosylated insulin-bound Concanavalin A with a suitable polymer that is permeable to both glucose and insulin, the glucose influx and insulin efflux would be controlled by the encapsulation membrane (Kost & Langer, 2012; Jeong, Kim, Eenink, & Feijen, 1984; Jeong, Kim, Holemberg, & McRea, 1985).

Another example of a closed-loop system is the urea-responsive strategy. Heller and Trescony (1979) had the idea of using immobilized enzymes to alter local pH and thus cause changes in polymer erosion rates. The proposed system is based on the conversion of urea to NH_4HCO3 and NH_4OH by the action of urease, increasing the pH, in a polymer that is subjected to increased erosion at high pH. Examples of polymers used are partially esterified copolymer of methylvinylether and maleic anhydride; *N*-hexyl half ester of methylvinylether and maleic anhydride; 4-carboxyacrylanilide copolymerized with methacrylate; and free radically cross-linked *N,N*-methylenebisacrylamide (Kost & Langer, 2012).

Nowadays, a great number of responsive systems displaying closed-loop characteristics are being developed, using the different interactions/reactions with the material of the system, with the aim of controlling drug delivery. The evolution can be observed by using antibody interactions and molecular biology.

According to this classification, it is clear that the strategies for the development of therapeutic systems aim for the capability of retaining high loads of active agents and selectively releasing them in targeted sites, improving therapy.

With the different ways to control drug release and the elevated degree of sophistication of many of these systems and active agents under current development, it is possible to control the release of attached or encapsulated drugs only at targeted sites in response to specific endogenous or externally applied stimuli.

References

Bruschi, M. L., Jones, D., Panzeri, H., Gremião, M. P. D., De Freitas, O., & Lara, E. H. G. (2007). Semisolid systems containing propolis for the treatment of periodontal disease: In vitro release kinetics, syringeability, rheological, textural, and mucoadhesive properties. *Journal of Pharmaceutical Sciences*, *96*(8), 2074–2089.

Bruschi, M. L., Jones, D., Panzeri, H., Gremião, M. P. D., De Freitas, O., Jones, D. S., et al. (2008). Precursor system of liquid crystalline phase containing propolis microparticles for the treatment of periodontal disease: Development and characterization. *Drug Development and Industrial Pharmacy*, *34*, 267–278.

Chan, A., Orme, R. P., Fricker, R. A., & Roach, P. (2013). Remote and local control of stimuli responsive materials for therapeutic applications. *Advanced Drug Delivery Reviews*, *65*, 497–514.

Ciach, T., & Moscicka-Studzinska, M. (2011). Buccal iontophoresis: An opportunity for drug delivery and metabolite monitoring. *Drug Delivery Today*, *16*(7/8), 361–366.

Edelman, E., Kost, J., Bobeck, H., & Langer, R. (1985). Regulation of drug release from polymer matrices by oscillating magnetic fields. *Journal of Biomedical Materials Research*, *19*, 67.

Farmer, T. G., Jr., Edgar, T. F., & Peppas, N. A. (2008). The future of open and closed-loop insulin delivery for diabetes mellitus. *Journal of Pharmacy and Pharmacology*, *60*(1), 1–13.

Heilmann, K. (1978). *Therapeutic systems—Pattern-specific drug delivery: Concept and development*. Stuttgart: Thieme.

Heller, J., & Trescony, P. V. (1979). Controlled drug release by polymer dissolution. II. An enzyme mediated delivery system. *Journal of Pharmaceutical Sciences*, *68*, 919–921.

Jadoul, A., Bouwstra, J., & Préat, V. (1999). Effects of iontophoresis and electroporation on the stratum corneum. Review of the biophysical studies. *Advanced Drug Delivery Reviews*, *35*, 89–105.

Jain, K. K. (2008). *Drug delivery systems*. Totowa: Humana Press.

Jain, R. K. (2012). Delivery of molecular and cellular medicine to solid tumors. *Advanced Drug Delivery Reviews*, *1*(64) (Suppl), 353–365.

Jeong, S. Y., Kim, S. W., Eenink, M. J. D., & Feijen, J. (1984). Self-regulating insulin delivery systems I. Synthesis and characterization of glycosylated insulin. *Journal of Controlled Release*, *1*, 57–66.

Jeong, S. Y., Kim, S. W., Holemberg, D., & McRea, J. C. (1985). Self-regulating insulin delivery systems. III. In vivo studies. *Journal of Controlled Release*, *2*, 143–152.

Jones, D. S., Bruschi, M. L., De Freitas, O., Gremião, M. P. D., Lara, E. H. G., & Andrews, G. P. (2009). Rheological, mechanical and mucoadhesive properties of thermoresponsive, bioadhesive binary mixtures composed of poloxamer 407 and Carbopol 934PR designed as platforms for implantable drug delivery systems for use in the oral cavity. *International Journal of Pharmaceutics*, *372*, 49–58.

Kost, J., & Langer, R. (2012). Responsive polymeric delivery systems. *Advanced Drug Delivery Reviews*, *64*, 327–341.

Kost, J., Wolfrum, J., & Langer, R. (1987). Magnetically enhanced insulin release in diabetic rats. *Journal of Biomedical Materials Research*, *21*, 1367.

Ottenbrite, R. M., & Kim, S. W. (2002). *Polymeric drugs and drug delivery systems*. Boca Raton: CRC Press.

Pereira, R. R. A., Godoy, J. S. R., Svidzinski, T. I. S., & Bruschi, M. L. (2013). Preparation and characterization of mucoadhesive thermoresponsive systems containing propolis for the treatment of vulvovaginal candidiasis. *Journal of Pharmaceutical Sciences*, *102*(4), 1222–1234.

Saslawski, O., Weigarten, C., Benoit, J. P., & Couvreur, P. (1988). Magnetically responsive microspheres for pulsed delivery of insulin. *Life Sciences*, *42*, 1521.

Shim, M. S., & Kwon, Y. J. (2012). Stimuli-responsive polymers and nanomaterials for gene delivery and imaging applications. *Advanced Drug Delivery Reviews*, *64*, 1046–1059.

Main mechanisms to control the drug release

4

4.1 Introduction

Drug delivery covers a very broad range of techniques used to get therapeutic agents into the human or animal organism. The strategies used to control the drug release from pharmaceutical systems are based on biological, physicochemical, and mathematical principles.

This field is interdisciplinary and has the objective of proper administration of bioactive agents to achieve a desired clinical response in the body, enhancing the drug therapy. Enhanced product safety, improved patient convenience, and compliance resulting from controlling the drug delivery are achieved by designing, and developing technologies that modify the temporal and spatial drug release profile.

Controlled release systems are developed with dependence on the active agent of interest, and aim to control drug exposure over time, assist the drug in crossing physiological barriers, shield the drug from premature elimination, and guide the drug to the desired site of action while minimizing drug exposure elsewhere in the organism. Therefore, the patient compliance to the therapeutic increases. In addition, it may have an addition of commercial value to marketed drugs by extending patent protection, and the variability of performance of drug products is reduced (Siegel & Rathbone, 2012).

There are several mechanisms by which the drug release can be controlled in a system. They are dependent on the particular application and may act simultaneously or at different stages of a process of delivery. In the development and manufacture of controlled release systems, it is important to understand these mechanisms. A controlled source of energy is necessary for the system to control the drug release at the proper schedule.

Moreover, there are chemical and biological mechanisms to control the drug release spatially. Basically, they are dissolution, diffusion, osmosis, partitioning, swelling, erosion, and targeting. It is common for a system or device to present more than one of them. The classification of controlled drug delivery systems regarding the mechanism of release is based on the main mechanism.

4.2 Dissolution

The active agents are from nature (animals, plants, or minerals) or synthetic, and most drug molecules form crystals at room temperature. The human or animal

Strategies to Modify the Drug Release from Pharmaceutical Systems. http://dx.doi.org/10.1016/B978-0-08-100092-2.00004-7

organism possesses a great amount of water and hydrophobic compounds (e.g., lipids), constituting the environment in which the drug needs to act. Therefore, it is very important to understand the concept of and relationship between some physicochemical properties and their importance to the drug release control.

The crystalline solids are characterized to contain highly ordered arrays of molecules and atoms held together by non-covalent interactions. They have proper morphology, particle size, and solvation/hydration depending on their processing conditions (Florence & Attwood, 2006). Therefore, their physicochemical characteristics, such as solubility, can be different according to the processing and crystallization. In some cases, particles of an active agent can be processed into an amorphous, glassy form. These forms have differing thermodynamic stabilities, and interconversion between solid forms can occur during storage and after administration. These changes can alter the solubility, stability, and biological activity of the agent.

The human or animal organism consists mainly of water, and the drug needs to be dissolved in this fluid to be released from the system and exert the therapeutic effect. Therefore, the dissolution is a very important mechanism that can control the drug delivery.

Dissolution can be defined as a process by which molecules of a solute (such as an active agent) are dissolved in a solvent vehicle. The dissolution process of an active agent involves the transfer of drug molecules or ions from its solid phase to the surrounding medium (generally water, tissue or, in some cases, polymer). The solution is said to be saturated when it contains a solute at the limit of its solubility, considering the conditions of temperature and pressure. Under certain circumstances, supersaturated solutions may be prepared and are characterized to contain the solute in solution above its normal solubility limit.

The rate at which the drug dissolves from the solid may be used to predict the drug release rate from the therapeutic system. The higher the solubility, the more rapid the rate of dissolution when no chemical reaction is involved (Florence & Attwood, 2006).

The dissolution process occurs when the solvating medium surrounding a solid drug particle is not saturated and can be controlled by the solvating medium, surface area of the solid, thickness of the boundary layer, and the diffusion coefficient (Figure 4.1). When the solid is placed in contact with the dissolution medium, the surface of solid contacts the medium and the molecules are solvated and removed from the solid. The solvated molecules (solute) increase in the medium, increasing the concentration of solute (C_s). This increase of solute concentration produces a boundary layer around the solid where the dissolution medium tends to be saturated and the dissolution rate decreases. The removal of the boundary layer causes the renewing of the solvent, increasing the dissolution rate. Thus, the difference between the solute concentration at the inner portion of the boundary layer (C_1) and the outer boundary layer (C_2) predicts the dissolution rates. When the difference is big, the dissolution rate is greater. However, when the difference is small, the dissolution rate is lower. The thickness of the boundary layer (l) is also important and conversely proportional to the dissolution rate.

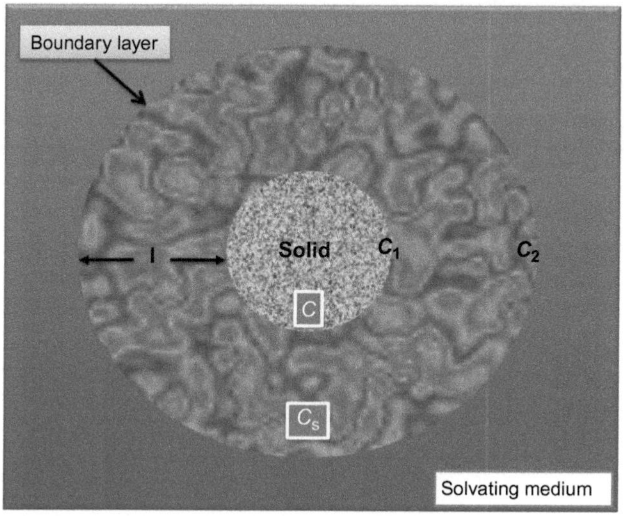

Figure 4.1 Dissolution process and rate: C_s is the concentration of solute; C is the concentration of solid; l is the thickness of the boundary layer; C_1 is the concentration of solute in the inner boundary layer; C_2 is the concentration of solute in the outer boundary layer.

Therefore, the dissolution process is mathematically described as

$$\frac{dC}{dt} = D \cdot A(C_s - C) \tag{4.1}$$

where the dissolution rate (dC/dt) is dependent on the diffusivity coefficient of solute (D), the surface area of solid (A), and the difference of solid solubility (C) and solute concentration (C_s).

Moreover, considering the thickness of the boundary layer and the difference of solute concentration in the medium (inner and outer boundary layers), the dissolution rate may be described as

$$\frac{dC}{dt} = \frac{D \cdot A}{l} \cdot (C_1 - C_2) \tag{4.2}$$

where the dissolution rate (dC/dt) is dependent on the diffusivity coefficient of solute (D), the surface area of solid (A) and the difference of the solute concentration at the inner (C_1) and outer (C_2) boundary layer, and is conversely proportional to the boundary layer thickness (l).

It is very important to note that the dissolution rate depends on solid particle size, agitation and viscosity of dissolution medium, and temperature. In this context, solubility is a thermodynamic property of an active agent and a medium. However, the dissolution rate is a kinetic property. The solubility coefficient of the drug in the medium decreases with the increasing melting point of the drug, and increases with the increasing chemical compatibility of the drug with the surrounding medium.

4.3 Partitioning

The drug partitioning in the medium is an important characteristic that influences the control of the drug release. Delivery systems can be constituted by one or more groups of materials displaying different affinities or polarities and leading the active agent to the organism.

Therefore, the drug molecules often find an interface between two materials (e.g., hydrophilic and hydrophobic polymers and oils) or phases (e.g., blood, water, and biomembranes). This movement of drug molecules from one phase to another is dependent on their relative concentrations (or chemical potentials) and their affinities for each phase.

This affinity can be measured by the ratio of drug solubilities in the two phases and is defined as partition coefficient

$$P = \frac{C_o}{C_w} P \tag{4.3}$$

where C_o is the concentration of the drug in the oil or hydrophobic phase and C_w is the concentration in the aqueous or hydrophilic phase.

Molecularly, the partition coefficient prescribes the relative frequency with which a molecule moves into one medium compared to the other, at the interface (Figure 4.2).

Some examples are illustrated in the literature (Siegel & Rathbone, 2012; Florence & Attwood, 2006). Active agents partitioning between aqueous and lipid biophases, such as drugs of high-lipid solubility, are suitable for entry into the *stratum corneum*. On the other hand, they have poor in-water solubility and their water/lipid partition coefficient is too low. Thus, they will not partition efficiently into the viable epidermis, and the drugs will be detained in the *stratum corneum*. Absorption into capillaries might then occur at an unacceptably low rate. Many systems are composed of polymers with different polarities, and the active agent will partition between these

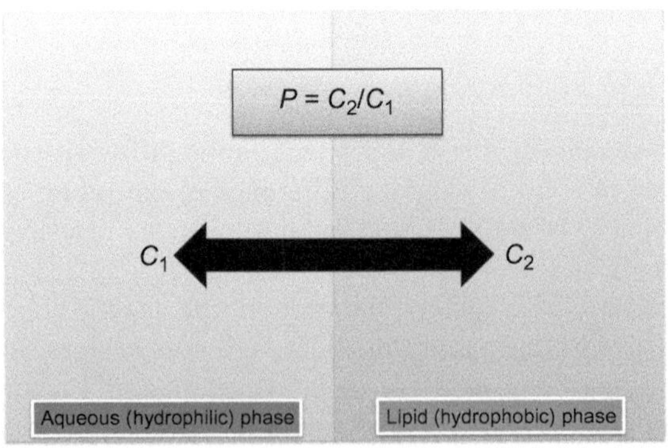

Figure 4.2 Schematic process of drug partitioning: C_1 is the drug concentration in the hydrophilic phase; C_2 is the drug concentration in the hydrophobic phase; P is the partition coefficient.

molecules. For example, it is possible to formulate polymeric systems containing micelles with hydrophobic cores and hydrophilic coronas; hence, they are soluble in the aqueous environment of the organism. Hydrophilic drugs will partition in the coronas and hydrophobic drugs will partition in the cores. The first ones are retained for a short period and the second ones are released for extended periods of time.

4.4 Diffusion

Diffusion is defined as the process of mass transfer of individual molecules of a substance from one part of a system to another, carried by random molecular motions, and is associated to forces such as concentration gradient. It is the process by which a concentration difference is reduced by a spontaneous flux of matter. This mass transfer at the system/water interface plays an important role in diffusion.

To explain diffusion, Crank (1975) gives the example of the classic experiment of a tall cylindrical vessel having its lower part filled with colored solution (e.g., iodine solution). When a column of clear water is carefully and slowly poured on top, no convection currents are set up and the colored part is separated from the clear by a sharp, well-defined boundary. However, over time, it is found that the upper part becomes colored, with the color getting fainter towards the top, while the lower part becomes correspondingly less intensely colored. As a consequence, the whole solution appears uniformly colored after sufficient time. In other words, the transfer of dye molecules from the lower to the upper part of the vessel took place in the absence of convection currents. Therefore, the solute is said to have diffused into the water.

The mass transfer from a solvent or solute is the base of many important phenomena of pharmaceutical areas. It is necessary for the diffusion through the biological membranes to the active agent to be absorbed and eliminated from the body and to reach the target site in a determined tissue or cell. Moreover, the drug release from a system can be controlled by this process.

All molecules undergo random and repeatedly collisions with other molecules due to thermal or Brownian motion. Over time, the movement of the molecule from one point to another is the result of a multitude of such random steps. Considering a simple solution, the solute will spontaneously diffuse from a region of high concentration to one of low concentration. Macroscopically, the independent random walks taken by a large number of drug molecules lead them from regions of higher concentration to regions of lower concentration.

In diffusion, the mass transfer is a kinetic process that occurs in systems that are not in equilibrium. The molecular diffusion is driven by the gradient of the solute concentration, considering the solution divided into volume groups with individual solutes, without preference for motion in any particular direction. The group of the molecules in this element may be considered to be moving in one determined direction. In an adjacent group, the same fraction of solutes may be moving in the reverse direction. If the concentration in the first volume group is greater than that in the second, the overall effect is that more particles are leaving the first element for the second, and hence there is a net flow of solute in the direction of the second group (less concentrated).

The random molecular motions are also the cause of heat transfer by conduction, and there is an obvious analogy between the two processes. In 1855, Fick verified that the mathematical equation of the basic Fourier principle of heat flow, developed in 1822, could be applied to mass transfer. He first puts diffusion on a quantitative basis.

Considering the scheme of Figure 4.3, it is possible to say that the mass transfer from point 1 to point 2 is governed by the concentration gradient and the distance. The amount of mass (M) flowing through a unitary transversal surface (S), in a time t, is defined as flux:

$$J = \frac{\mathrm{d}M}{S \cdot \mathrm{d}t} \tag{4.4}$$

This is Fick's first law, where the molecule flux is proportional to the gradient of concentration

$$J = -D \cdot \frac{\mathrm{d}C}{\mathrm{d}x} \tag{4.5}$$

$$\frac{\partial C}{\partial t} = -D \cdot \frac{\partial^2 C}{\partial x^2} \tag{4.6}$$

where J is the rate of transfer per unit area of the section, C is the concentration of the diffusing substance, x is the distance between the two points, and D is called the diffusion coefficient.

When the solution is diluted, the diffusion coefficient D can reasonably be taken as a constant. However, in other situations (e.g., diffusion in high polymers), it depends very markedly on concentration (Crank, 1975).

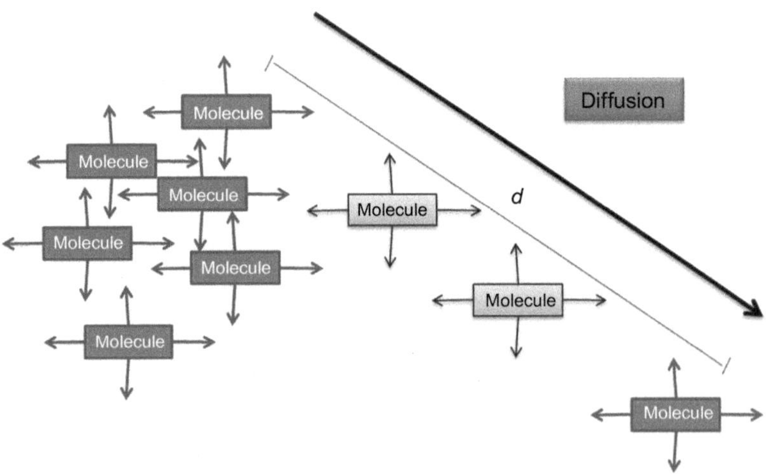

Figure 4.3 Diffusion process of molecules through a solvent: (d) distance of molecule movement.

Assuming diffusing spherical particles or molecules, the relationship between its radius and diffusion coefficient is given by the Stokes–Einstein equation:

$$D = \frac{RT}{6\pi\eta r N_{\mathrm{A}}} \tag{4.7}$$

where R is the general constant of gases ($8.314\,\mathrm{J\,mol^{-1}\,K^{-1}}$), T is the thermodynamic temperature (K), η is viscosity (Pa s), r is the radius of the particle or molecule, and N_{A} is Avogadro's number ($6.02\times10^{23}\,\mathrm{mol^{-1}}$).

Observing the previous equation, it is possible to verify that the diffusion coefficient (D) has unities of area and time and depends on parameters related to the solvent and the molecule. The negative sign in the equation of Fick's first law arises because diffusion occurs in the direction opposite to that of increasing concentration.

When there are macromolecules, the equation is modified to consider the form factor of the molecule:

$$D = \frac{RT}{6\pi\eta r N\mathrm{A}} \cdot \sqrt[3]{\frac{4\pi N\mathrm{A}}{3MV}} \tag{4.8}$$

where M is the molecular mass and V is the specific partial volume (which presents values very similar to the volume occupied by one gram of polymer).

The diffusion coefficient D is not properly a constant, but it is a measure of the molecule's mobility in the medium and depends on temperature, pressure, properties of the solvent, chemical nature solute, and the diffusant concentration, mainly when it is in higher concentrations. Large molecules diffuse more slowly than small ones, and that diffusion should be slowed in viscous liquids. Moreover, the intensity of thermal agitation drives Brownian motion.

The diffusion occurs in the direction of concentration decreasing. Therefore, the flux is always a positive amount and the movement will stop when the gradient of concentration disappears. Moreover, the average (actually, root mean squared) distance that solute molecules travel by diffusion is proportional to the square root of time. Conversely, the typical time required to diffuse over a particular distance is proportional to the square of that distance and inversely proportional to the diffusion coefficient.

When the system is a polymeric matrix that does not flow like a liquid, bulk viscosity is not the parameter to use in predicting the mobility of the active agent. However, the matrix is a dynamic fluctuating structure, and the diffusion coefficient may be considered as a measure of the degree that these fluctuations accommodate random motion of the diffusing molecule. The free volume (nanovoids) of matrix changes in size as the temperature varies, and a void can became large enough for diffusing molecule to move into or through it. Therefore, the increasing of the molecular radius of a molecule and of the matrix's density decreases the diffusion coefficient. At a critical density, often associated with the medium's glass transition temperature, T_{g}, free volume becomes so sparse that the diffusion coefficient drops by several orders of magnitude.

The composition of matrix influences in the diffusion coefficient due to the free volume depends on its composition. A simple matrix (homogeneous) shows an increase of free volume as the difference between the ambient temperature and the temperature of glass transition increases. A heterogeneous matrix, prepared by copolymerization or blending, can display suitably averaged free volumes and mobility properties. Free volume can also be increased substantially by sorption of small molecules, such as water.

The coefficient of permeability (P) is defined as the velocity of diffusion of the diffusant through the mean and is equivalent to the distance that the drug travels by time unit, considering the similarity with the linear velocity. Therefore, P possesses time and distance units

$$P = \frac{DK}{h} \tag{4.9}$$

where K is the coefficient of partition and h is the thickness of the membrane or the distance traveled by the active agent.

The statement expressed mathematically in Fick's first law is in general consistent only for an isotropic medium displaying symmetry. When the medium is anisotropic, the structure and diffusion properties in the neighborhood of any point can be different relative to all directions, and the diffusion properties depend on the direction in which they are measured (Crank, 1975). However, in pharmaceutical areas, the unidirectional movement is enough to describe the majority of cases.

Moreover, the aforementioned equation of Fick's first law considers the element of volume for a plane geometry (e.g., thick slab). Equivalent equations of Fick's first law are available to describe drug transport through different geometries (Crank, 1975):

(1) Cylinders

$$\frac{\partial C}{\partial t} = \frac{1}{r}\left\{\frac{\partial}{\partial r}\left(rD\frac{\partial C}{\partial r}\right) + \frac{\partial}{\partial \theta}\left(\frac{D}{r}\frac{\partial C}{\partial \theta}\right) + \frac{\partial}{\partial z}\left(rD\frac{\partial C}{\partial z}\right)\right\} \tag{4.10}$$

considering an element of volume of a cylinder of sides dr, r, dθ, and dz.

(2) Spheres

$$\frac{\partial C}{\partial t} = \frac{1}{r^2}\left\{\frac{\partial}{\partial r}\left(Dr^2\frac{\partial C}{\partial r}\right) + \frac{1}{\sin\theta}\frac{\partial}{\partial \theta}\left(D\sin\theta\frac{\partial C}{\partial \theta}\right) + \frac{D}{\sin^2\theta},\frac{\partial^2 C}{\partial \varphi^2}\right\} \tag{4.11}$$

considering an element of volume of a sphere of sides dr, r, dθ, r, sinθ, d$\emptyset$.

The environmental conditions that constrain the carrier system are captured by initial conditions and solutions to Equations (4.5) and (4.6) that have been obtained for a variety of initial and boundary conditions and described by Crank (1975).

Fick's second law of diffusion is the basis for the majority of mathematical models of diffusion processes. It aims to determine the rate of concentration variation of diffusing at a determined point of a system (Peppas & Narasimhan, 2014). This law is the equation of mass transport that emphasizes the variation of concentration in function of time in a determined place, and not the mass diffusion through a unit of area by time:

$$\frac{\delta C}{\delta t} = D\left(\frac{\delta^2 C}{\delta x^2} + \frac{\delta^2 C}{\delta y^2} + \frac{\delta^2 C}{\delta z^2}\right) \tag{4.12}$$

where C is the concentration of the diffusant; t is time' D is the diffusion coefficient and x, y, and z are the directions.

Thus, the concentration in function of time in a determined region is proportional to the alteration in the concentration gradient in the same region of the system.

Moreover, diffusion is an efficient means of mass transport over short distances, but its effectiveness decreases over longer distances. Considering the reservoir and monolithic systems, it is necessary to understand the process of drug diffusion from these different conditions. In a reservoir system the active agent is contained by a controlling membrane, while in a monolithic system the active agent is homogeneously dispersed in the polymeric matrix.

The molecular diffusion through polymers and synthetic membranes is an efficient, secure and simple strategy to reach the control release of several active agents. It is said that a therapeutic system or device acts by diffusion when this phenomenon occurs in some phase of release, and represents a decisive step in the total release of the drug.

In reservoir systems, the purpose of the membrane is to mediate diffusion of the drug. The drug first partitions into the membrane from the reservoir and then diffuses to the other side of the membrane, where it is taken up by the receiving medium. The release steps of this phenomenon are illustrated in Figure 4.4. The drug needs to diffuse along the reservoir, dissolve, or partition between the carrier fluid and the membrane, diffuse through the membrane, partition between the membrane and the elution medium (external) and, finally, transport from the system surface.

While the reservoir is saturated, a constant concentration gradient of the drug is maintained in the membrane, the rate of drug flux is constant, and zero-order release is

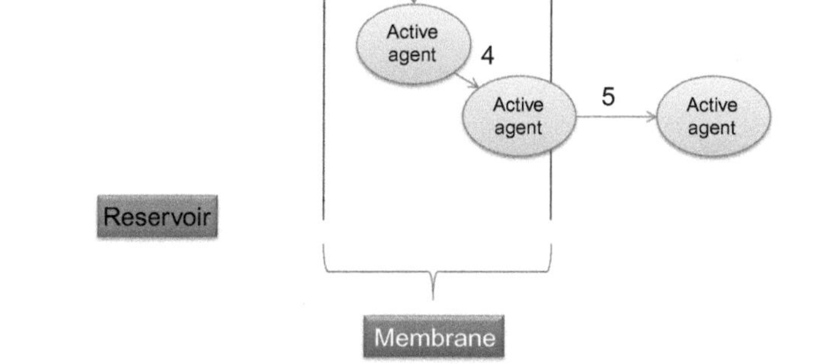

Figure 4.4 Schematic representation of steps of active agent release from a reservoir system: (1) diffusion along the reservoir; (2) dissolution or partitioning between the carrier fluid and the membrane; (3) diffusion through the membrane; (4) partitioning between the membrane and the external fluid medium; (5) transport from the system surface. The concentrations of active agent through the sites are (C_d) reservoir; (C_1) inner membrane; (C_2) outer membrane; (C_r) external medium.

achieved (Siegel & Rathbone, 2012). However, the concentration of active agent in the reservoir may fall below the saturation, indicating that the amount of active is not constant. In this case, the gradient across the membrane and the release rate both decay.

Another important factor for reservoir devices is the nature of the fluid inside the reservoir. When the drug is coarsely dispersed (suspension) in the fluid, the default of good contact with the membrane will conduce to inconstant and possibly non-reproducible release rates. Liquid vehicles have the possibility of good contact with the membrane due to the movement of the vehicles themselves. However, it is necessary to take precaution that the liquid does not flow to the exterior together with the active agent.

The mechanism of reservoir systems is simple and has the ability to produce zero-order release, making its use highly advantageous. On the other hand, it is a type of system that can be difficult to fabricate reliably, and dose dumping can occur due to pinhole defects and cracks in the membrane (Siegel & Rathbone, 2012).

Monolithic systems are characterized to be constituted with the active agent loaded directly into a matrix, acting as both a storage medium and a mediator of diffusion, and can avoid these problems of reservoir systems.

Considering the basic principle that the drug is loaded uniformly into the monolithic system, it is released under control of diffusion through the matrix or through aqueous pores. When the active agent is not dissolved (coarsely dispersed), it constitutes a separated phase that needs to dissolve in the matrix to be released. The previous dissolution in the matrix is necessary to perform the diffusion to the surface. In this case, the dissolution is a fundamental step before the diffusion. Therefore, the release steps of the active agent from monolithic systems can be different (Figure 4.5). The first step varies and depends on drug solubility in the matrix. When the drug is dissolved in the matrix, the diffusion occurs at the device surface. When it reaches this place, the partitioning of the drug between the matrix and the elution medium occurs, as well as the transport from the device surface.

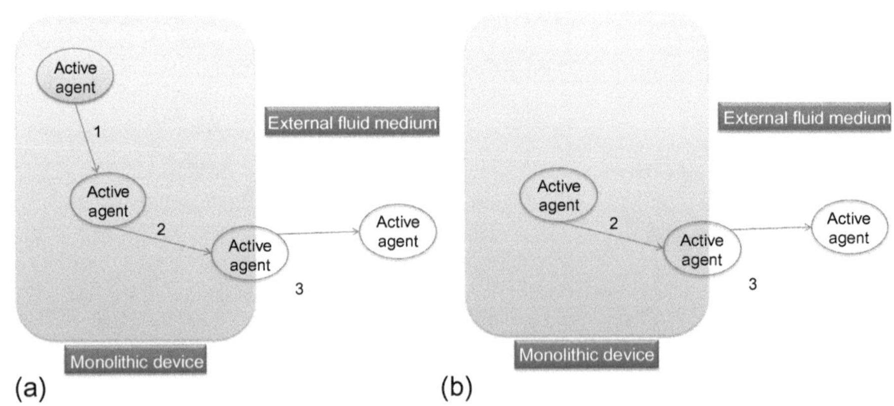

Figure 4.5 Schematic representation of steps of release from a monolithic system of an active agent dissolved (a) or not (b): (1) dissolution in the matrix; (2) diffusion through the matrix to the surface; (3) transport from the system surface.

Monolithic devices typically exhibit an initial burst of release from the surface, and the release rate decreases with the passing time. This phenomenon is a consequence of the distance of the drug from the surface. Contrary to what occurs with reservoir devices, in monolithic devices, the mean distance that the molecules of the active agent need to travel for diffusion before being released from the device varies with time as the active is being expelled from the device. Therefore, the geometry of the system has great influence on the drug release.

The geometries more commonly found in controlled release systems are planar, cylindrical and spherical. In addition, the geometric factor can be substantially reversed using special geometries such as hemisphere monoliths, specially coated wedges, cones, and sectioned cylinders. The goal of these special geometries is to create a crescent area in which the active agent dissolves in the matrix as the diffusion distance increases, obtaining constant release rates or near-zero-order release.

A drug that is deeper inside the monolith must diffuse to the surface because it has farther to travel, and the quadratic relation between distance and time becomes important. This effect occurs in planar monoliths, but it is even more prominent with cylinders or spheres, as the amount of drug available decreases with distance from the surface (Fan & Singh, 1989; Higuchi, 1961, 1962; Siegel & Rathbone, 2012).

The boundary layer (hydrodynamic diffusion layer) is the layer localized on the external surface of the system. This layer exerts an extra resistance to the diffusion and can restrict or slow the transport of the active agent from the device. When the active has low solubility, the phenomenon may be extreme and the boundary layer can become saturated with the active agent and stop the drug release from the system.

During the diffusion process, the environmental conditions can constrain the system and are captured by initial and boundary conditions. Solutions to equations of Fick's first law have been obtained for a wide variety of initial and boundary conditions (Crank, 1975).

4.5 Osmosis

The choice of the right drug and dosage over time is fundamental for effective therapy. Rate-controlled release systems possess the property of maintaining the active agent concentration within the body at an optimum level, reducing the risk of side effects and poor therapeutic activity.

Over the past 50 years, the application of one of the most fundamental principles of biology, osmosis, in a technical device, has enabled the control of the drug release rate. Drug delivery systems based on the principles of osmosis have been proposed as an effective manner to control the release rate.

They are well-established, trusted systems that enable a wide range of applications, from oral dosage forms providing less than 24 h of release to implantable devices that control precisely the release of the active agent over periods of months, years or more. Moreover, the osmosis principle can be used together with other mechanisms to control the drug release rate.

Osmosis may be defined as the action in which the solvent is transferred through a semipermeable membrane to dilute a solution containing solute and solvent. When two solutions of different concentrations are separated by a semipermeable membrane that is permeable to the smaller solvent molecules but not to the large solute molecules, then the solvent will tend to flow through the semipermeable membrane from the less concentrated to the more concentrated solution. It is an effort to equalize concentrations of the impermeable solutes on both sides of the membrane (Figure 4.6).

Considering biological organisms, the solvent is water. The transport of water and other molecules across biological membranes is essential to many processes in living organisms. Osmotic pressure is the energy that drives the process and controls the drug release from therapeutic systems.

However, for active agents, the nature of water transport may differ from this simple concept. The concentration of solutes (also called osmolytes) and the thermodynamic compatibility of water with these osmolytes determines the chemical potential, which determines the flux of water through the membrane together with the concentration gradient of water. When the osmolytes are small molecules (e.g., salts), osmotic pressure is mainly determined by osmolyte concentration, but when the osmolytes are polymers, osmotic pressure is determined jointly by polymer concentration and polymer/water compatibility (Siegel & Rathbone, 2012).

The greater the gradient in chemical potential, the greater the rate of transport of solvent through the membrane. Considering the relationship between chemical potential and osmotic pressure, and the fact that the water molecules are in motion and in contact with each other as a result of hydration of the membrane, the rate of osmotic flow across a unit area of the membrane is determined by (Amkraut, Eckenhoff, & Nichols, 1990; Siegel & Rathbone, 2012):

- The concentration and nature of osmolytes on both sides of the membrane
- Temperature
- The hydraulic permeability of the membrane, which can be determined by measuring water flow when hydrostatic pressure is applied across the membrane

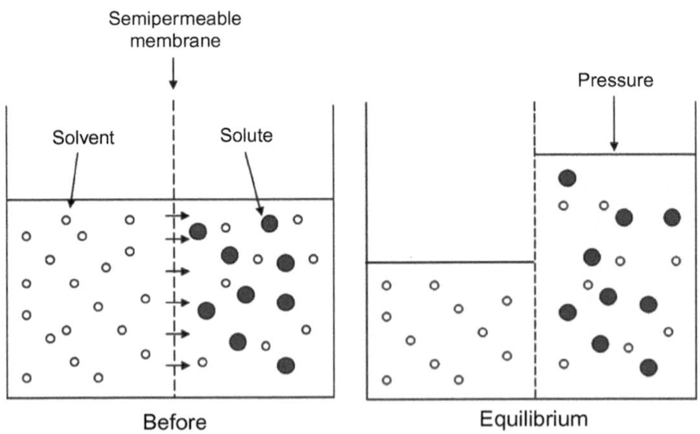

Figure 4.6 Diagram of principle of osmosis.

Theeuwes (1975), studying the concept of an elementary osmotic pump, described that the rate of water transport through the membrane can be written as follows:

$$\frac{\mathrm{d}V}{\mathrm{d}t} = \left(\frac{A}{h}\right)\mathrm{Lp}\left(\sigma\Delta\pi - \mathrm{Jp}\right) \tag{4.13}$$

where $\mathrm{d}V/\mathrm{d}t$ is the volume flow of solvent through the membrane, A is the cross-sectional area for transport, h is the membrane thickness, Lp is the hydraulic permeability of the membrane, σ is the reflection coefficient, $\Delta\pi$ is the osmotic pressure difference across the membrane, and Jp is the hydrostatic pressure difference across the membrane.

It is important to emphasize that the osmotic water flow is reduced when the membrane is partially permeable to the osmolytes. As water flows into a device containing osmolytes, it dilutes the osmolytes, lowering the osmotic pressure, unless new osmolytes are introduced, for example, by dissolution.

Osmotic pressures of saturated aqueous solutions of selected pharmaceutical solutes (e.g., sodium chloride, potassium chloride, potassium sulfate, mannitol, heptahydrated sodium phosphate dibasic, fructose, sucrose, and dextrose) are normally higher than 30 atm (Mathiowitz, 1999). Hydrostatic pressure inside most osmotic drug delivery systems is generally less than 1 atm, although some systems may attain pressures as high as several atmospheres (Theeuwes & Ayer, 1978). Therefore, the hydrostatic pressure differential ($\Delta\pi$» &Δp) is negligible and the equation can then be written as

$$\frac{\mathrm{d}V}{\mathrm{d}t} = \left(\frac{A}{h}\right)k\Delta\pi \tag{4.14}$$

where $k = \mathrm{Lp}\cdot\sigma$, and it can be taken as the effective permeability of the membrane.

Based on this principle, controlled release systems imbibe water from the body through a semipermeable membrane into an osmotic material, which swells, resulting in slow and controlled delivery of the active from the formulation. From this mechanism the osmotic pumps are introduced, as shown in the next chapters.

The coating of individual drug particles with semipermeable polymers can also control the drug release. When these particles are exposed to water (e.g., of gastric fluid), the solvent crosses the polymer coatings and dissolves the active agent, leading to a gradient in solute concentration that drives even more water inside. To accommodate this, the coating must expand, and wall stresses are developed. According to the osmotic driving force, the coating ruptures, and the active is released. Thus, using different coating thicknesses, particles can be programmed to burst at different times. The original time-release capsules were based on this principle. Moreover, tablets or particles coated with a semipermeable membrane that possesses sparse but well-distributed aqueous pores are other examples of the utilization of the osmosis principle to control the release rate. The membrane pores can be created by excipients blended into the membrane, which dissolve upon exposure to water. The pores constitute spaces through the water flows across the semipermeable parts of the membrane, and displace the dissolved active agent through the aqueous pores into the release medium (Siegel & Rathbone, 2012).

In comparison with drug delivery systems based on diffusion and erosion, osmotic systems tend to be less volume efficient and more complex in design. However, osmotic systems also tend to provide better zero-order delivery and to deliver a greater percentage of the drug loading at a zero-order delivery rate (Mathiowitz, 1999).

4.6 Swelling

Many materials display the characteristic of swelling when in contact with water. This is due to their hydrophilic behavior and the interaction between water and their molecules. Polymers are the main materials employed for preparation of controlled drug release systems, and their polymer chains can organize in a three-dimensional fashion. When a polymer network is surrounded by water, the network expands and chemical or physical bonds are formed (Figure 4.7). Compared to osmosis, the swelling process is very similar. Water enters the polymer relatively rapidly, while dissolution of the polymer into water, if it occurs, is comparatively slow because of the need for polymer chains to disentangle (Siegel & Rathbone, 2012). This expansion of volume and the consequent emergence of greater spaces between the polymeric chains can be used to control the release of active agents from polymeric systems (matrix or reservoir).

In practice, the respective significance of the swelling can be macroscopically observed from the morphology of the matrix or membrane (in the case of reservoir systems). In the early stage of drug delivery, the uptake of water by a polymer system increases the thickness and volume. The successive polymeric chain disentanglement and the dissolution of the drug or fillers used counteracts this increase in thickness,

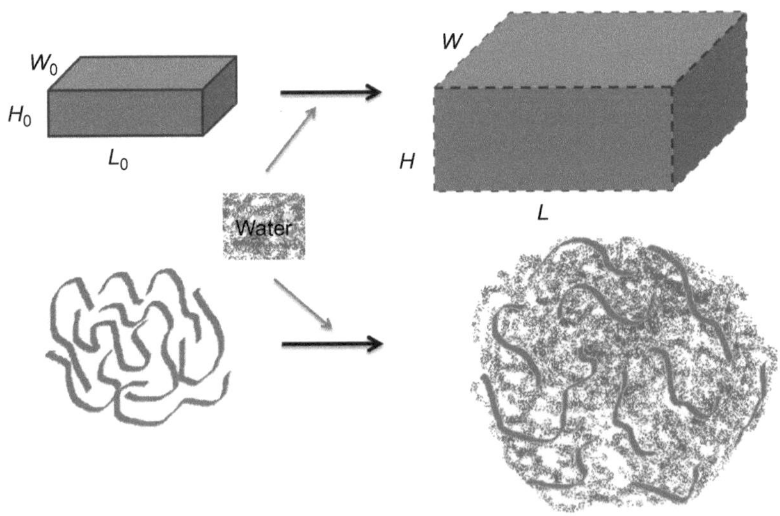

Figure 4.7 Schematic diagram of swelling process: (W_0) initial width; (H_0) initial height; (L_0) initial length; (W) final width; (H) final height; (L) final length.

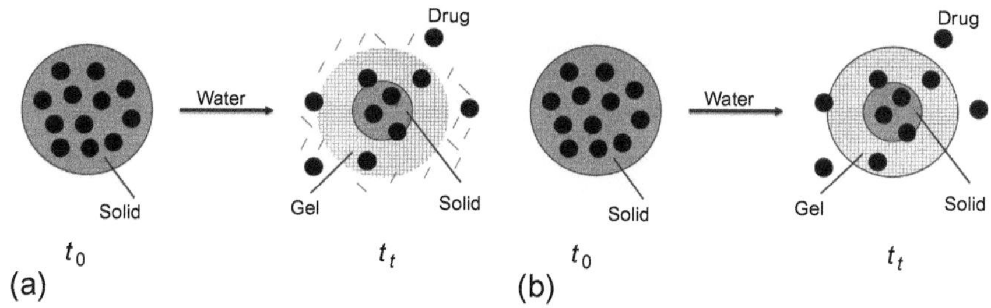

Figure 4.8 Swelling process during the time period of $t_0 - t_t$: (a) unlimited and (b) limited.

producing a diminution of the volume of the matrix. Finally, the matrix disappears when the entire polymer is swollen (Figure 4.8a). This type of system can be more opportunely called a "swellable-soluble matrix" (Colombo, 1993). In this case, swelling is often a prelude to polymer dissolution.

However, sometimes the swelling of material is limited and the matrix does not disappear (Figure 4.8b). This phenomenon can occur if water and the polymer are insufficiently compatible, if polymer chain length is sufficiently large, or if cross-links are introduced to form a polymer network (Siegel & Rathbone, 2012).

The driving force for the water uptake and swelling process is generally a balance of the forces of osmotic, electrostatic and entropy-favored dissolution of the polymer in water. Elastic forces are tailor-made in the hydrogel structure to control the entropy of the dissolution process (Omidian & Park, 2008). The polymer's hydrophilicity and the presence and density of cross-links between polymer chains will determine the extent of swelling. Thus, hydrophobic polymers show low affinity by water and hence do not swell significantly (Siegel & Rathbone, 2012). Therefore, the velocity of swelling will control the drug release.

Swelling controlled release systems are typically glassy polymers at room and body temperatures. It is possible to divide the swelling agents into chemical and physical. In the first group are the polymers that possess all chains cross-linked to each other by covalent bonds. During the synthesis of chemical polymeric material, a cross-linker is used to link the polymeric chains. When the precursor hydrophilic polymer is treated physically, a physical polymeric material is originated. Polymers should therefore contain functional groups capable of reacting with ions, or other functional groups like aldehydes (Colombo, 1993).

When in contact with water, the glassy polymer at the surface of the system relaxes to a configuration that is more compatible with water, and swells. This opens the way for water to intrude even further, and a moving front is often observed, separating a swollen outer layer from a dry inner core (Figure 4.9).

Cross-link content, ionic content and hydrophilic content are the three major elements of swelling process control. Controlled release drug delivery may take advantage of tailor-made swelling kinetics, which can be provided by different means, including a change in cross-link density (Omidian & Park, 2008). The relaxation of polymeric chains usually conduces to a glass-to-rubber transition.

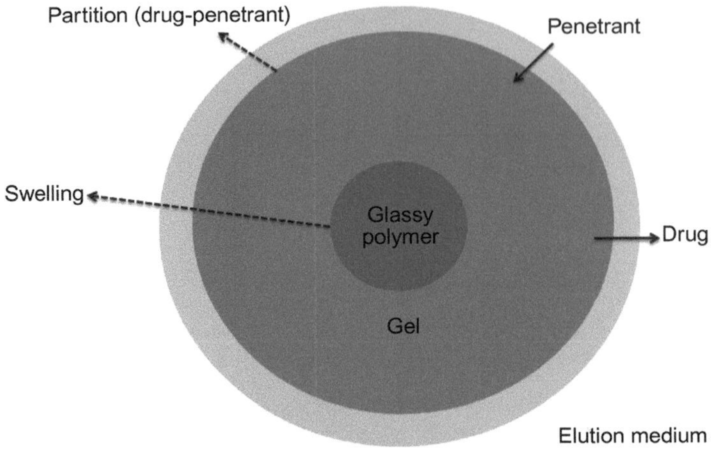

Figure 4.9 Schematic representation of phenomena that occur in swelling-controlled release systems.

The active agent, additives, proper selection of polymer, temperature and pH can modulate the swelling and the control of release rate. Reversible swelling and shrinking of hydrogels can also be induced by alternating these parameters with concomitant on/off patterns of drug release (Siegel & Rathbone, 2012).

4.7 Erosion

The polymers used in drug delivery systems controlled by diffusion possess a relatively passive function. They are carriers and delay the velocity with which the active agent is distributed to the targeting site. Some polymeric carriers are elaborated to play a more active function on the drug delivery process. These polymers suffer erosion when they suffer chemical reactions, releasing the active agent.

For implantable or injectable therapies, these systems are popular because they do not require retrieval after the active agent is fully released. They are classified into two categories (Figure 4.10):

(1) Systems of physical immobilization
(2) Systems of chemical immobilization

In the first category, the active agent is physically immobilized by the polymeric net, and it is released after the erosion of the net. These systems are also called erodible or biodegradable.

When the active agent is chemically linked to the polymeric structure or is itself part of the structure, the system is formed by chemical immobilization (second category).

The advantages of an erodible system are evident, especially for pharmaceutical applications. Because the system will be disappear in the site after being administered, it will not need invasive intervention to be removed. However, the erosion products

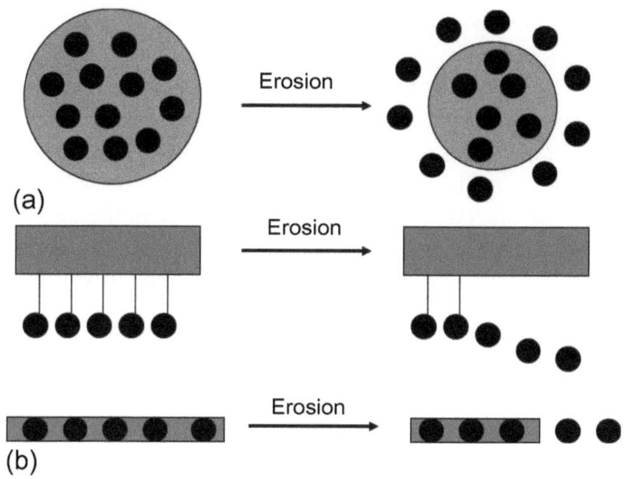

Figure 4.10 Drug delivery systems controlled by erosion: (a) physical immobilization; (b) chemical immobilization.

must be non-toxic and excretable or resorbable. Moreover, a system of pendant chains and of polydrugs offers the additional possibility of drug targeting and reducing the toxicity of the active agent.

The drug release from these systems is governed mainly by kinetic degradation of the appropriate link, which is specific for each system. Two limits of behavior in erodible systems (i.e., surface erosion and bulk erosion) need to be considered.

Surface erosion of the system occurs either when water invasion is slow or hydrolysis is rapid. When the polymer is hydrophobic, its chemical bonds susceptible to hydrolysis are protected from exposure to water in the interior of the polymeric system. Thus, hydrolysis with accompanying release of the active agent only occurs at or close to the surface. Surface erosion leads to the reduction of system dimensions with time (Figure 4.11).

Pure surface erosion is almost impossible to achieve, and diffusion of the drug out of a matrix may occur ahead of erosion. The drug itself may draw in water, and

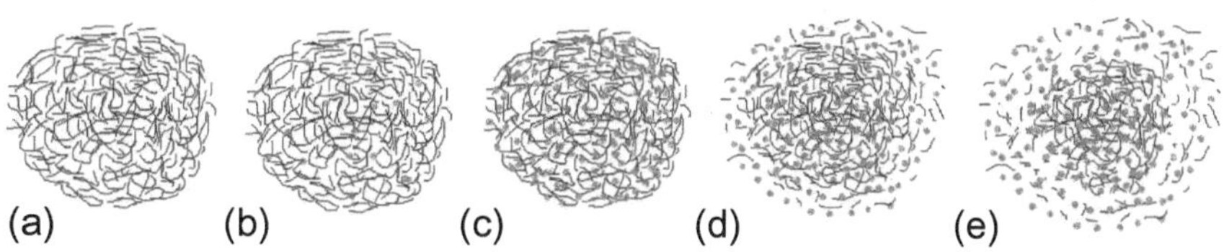

Figure 4.11 Diagram of stages of system surface erosion: (a) before contact and (b) just after contact with surrounding media; (c and d) active agent trapped in the outer layers of the system starts to be released into the surrounding media following erosion of the system surface; (e) remaining active agent is trapped in the delivery system, which reduces in size as time progresses and with the consequent polymeric erosion from the surface inwards, eventually resulting in all drug being released.

osmotic stresses (due also to small chain fragments) in the polymer can lead to fracture and uneven penetration. Hydrolysis of bonds is the main way that polymeric erosion occurs, and it may also be acid or base catalyzed, depending on the local concentration of proton donors and acceptors. An erodible system containing the active agent physically added can be prepared using the same way as those controlled by diffusion (i.e., reservoir or monolithic systems).

However, when the surrounding media (water) invades the system more rapidly than hydrolysis can occur, the water establishes its presence throughout the material, chain scission processes are initiated everywhere and bulk erosion occurs. It is characterized by a very slow hydrolysis initially, and initial scissions may endow chains with sufficient mobility that they migrate and form crystallites, which are less susceptible to hydrolysis. After a determined grade of hydrolysis has occurred, the process may accelerate. The formation of short chains may lead to an overall loss of polymer and increase in water concentration by diffusion and/or osmosis. Moreover, if chain scission results in the formation of acidic end groups and the scission process is acid catalyzed, then erosion will be autocatalytic.

Therefore, bulk erosion typically exhibits three stages, as shown in Figure 4.12. First, the drug is released from the system surface or from pores that are connected to the surface. The second stage is characterized as a latent stage in which a little degradation of polymer takes place, and the remaining active agent is trapped. Finally, in the last stage, the trapped active agent is released rapidly when the polymer disintegrates. This rapid degradation in the interior of the system is due to accumulation of autocatalytic erosion products, while leaching of these products leads to slower erosion at the surface. When this is true, thicker matrices may erode more rapidly than thinner matrices (Siegel & Rathbone, 2012).

Considering a reservoir system, the erosion of the controlling membrane can also significantly affect the release rate of the active agent. This allows a mechanism to get a wide range of distribution rates, when the erosion and release rates are coordinated. However, an unexpected loss of membrane integrity by erosion can lead to an undesirable release of the active agent.

A more practical reservoir system would be that in which the erosion process is slow during the release period, not affecting the release rate. However, with a very slow erosion, it is possible that the system will disappear from the site before the entire drug is released.

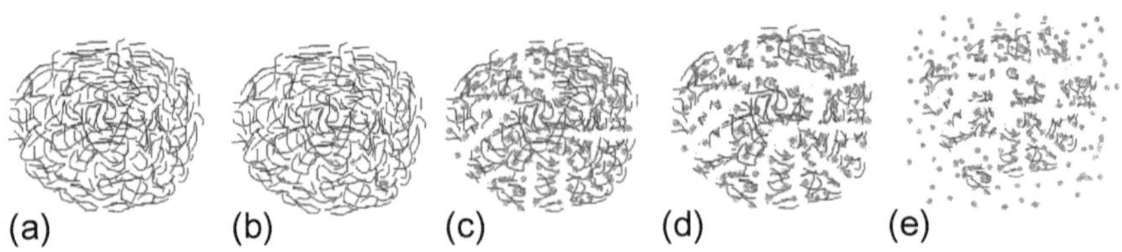

Figure 4.12 Schematic representation of the stages of bulk erosion: (a) before contact and (b) just after contact with surrounding media; (c) stage of drug release from the system surface and pores connected to surface; (d) latent stage of release; (e) polymer disintegration and rapid release of active agent.

A monolithic system prepared with an erodible polymer can release the active agent by erosion, diffusion, or both, depending on the relative velocity of two phenomena. The erosion of a polymeric system can occur from the surface to the interior (heterogeneous or surface erosion) or homogenously throughout the system (homogeneous erosion).

A polymer that displays surface erosion is desirable, and if the diffusion is not significant, the surface area remains constant and zero-order release takes place. The mechanical loss of system integrity is less probable, and the release rate will be independent from physical and chemical properties of the active agent.

The degradation mechanism more desirable in these systems is the hydrolytic cleavage of labile bonds. Hydrolysis of bonds may also be acid or base catalyzed, depending on the local concentration of proton donors and acceptors. The enzymatic degradation can trigger erosion by cleaving of polymer chains or cross-links into a polymer network (e.g., reducing agents that degrade polymers that include disulfide bonds).

The nature of the erosion mechanism can be provided by the hydrophilicity and morphology of the polymer. Hydrophobic polymers tend to exclude water from the interior of the system, and they will erode from the outer to inner portion of the system. On the other hand, hydrophilic polymers favor homogeneous erosion, and the crystallinity grade also has an influence on the process, once the crystalline regions of polymer tend to exclude water with more intensity than the amorphous region.

Depending on the erosion type of polymers, the polymers and the systems are classified in different groups. The most common erodible systems are based on poly(lactic acid) or poly(lactic acid-*co*-glycolic acid), although systems based on poly(ε-vinyl caprolactone), poly(ortho esters), polyanhydrides, polyphosphates, poly(phosphazenes) and pseudopoly(amino acids) have also been utilized or studied (Ottenbrite & Kim, 2002; Wang, Siahaan, & Soltero, 2005; Siegel & Rathbone, 2012; Wilson & Crowley, 2011).

4.8 Targeting

The requirements of strategies for construction of a therapeutic system must first be determined, focusing on the benefit of the enhancement of the active agent.

As discussed in Chapter 1, when the active agent can be administered at its preferred site of action and kept away from sites associated with toxicity, the therapy becomes safer. When the active agent is concentrated at the site of action rather than being dispersed throughout the body, its efficacy is also improved. With the increased safety and efficacy, the therapy becomes more acceptable and thus improves patient compliance.

4.8.1 Concepts and definitions

More than a century ago, Paul Ehrlich (1954) described an active agent that is aimed precisely at a disease site and that would not harm healthy tissues as a "magic bullet." However, at therapeutic concentrations, very few drugs bind solely to their intended therapeutic target. From these considerations, a concept of site-specific drug-delivery systems was formed, and a therapeutic agent would be attached to a carrier that would take it to the target (attached to the carrier via a targeting ligand) and release it at the target site (Petrak, 2005).

Drug targeting is defined as the ability of the active agent to accumulate in the target site selectively and quantitatively, independent of the site and methods of administration. The local concentration of the drug at the site(s) should be high, while its concentration in other non-target organs and tissues should be below certain minimal levels to prevent negative side effects. Thus, drug targeting is a very important mechanism of drug delivery control. A system with optimal targeting properties should contain structural features that specifically interact with the designated target, for example, a particular receptor utilizing ligand–receptor interactions (Vyas et al., 2001). Ideally, such a receptor should only be associated with a diseased organ or tissue (Petrak, 2005).

The "magic bullet" or similar words found in newspaper articles or public magazines have often fascinated general readers and scientists because these words in some way implicate remote, seeking, and selective destruction (Bae & Park, 2011).

The concept of drug targeting was initially used to describe the possible *in vivo* action of an anticancer drug conjugated to a monoclonal antibody that is specific to a surface antigen on target cancer cells. However, targeting can occur at the specific organ (e.g., brain to treat Parkinson's, Alzheimer's and Creutzfeldt–Jakob disease or targeting the lung in the treatment of cystic fibrosis), tissue (e.g., targeting site of inflammation or tumors), cellular (e.g., targeting to trastuzumab to human epidermal growth factor receptor-2 and cancer cells), subcellular compartment or organelle level (e.g., targeting to cytoplasm, proteins, receptors, mitochondria, DNA and gene therapy applications), or invading organisms (e.g., targeting viruses, parasites and bacteria), and a range of targeting systems were developed.

The term "targeted drug delivery" (or "drug targeting") used in drug delivery refers to predominant accumulation of the active agent within a target site, and it is independent of the method and route of drug administration. It is important to understand the difference for "targeted therapy" (or "targeting therapy"), which is frequently used in drug discovery and means specific interaction between an active agent and its receptor at the molecular level. Effective targeted drug delivery systems have four key requirements: retain, evade, target, and release (Mills & Needham, 1999).

4.8.2 Advantages

The advantages of drug targeting are evident:

- Drug protocols may be simplified.
- The drug quantity required to achieve a therapeutic effect may be greatly reduced, as well as the cost of therapy.
- Drug concentration in the required site can be sharply increased without negative effects on non-target places.

The route by which a system is administered can have a strong impact on the rate of drug delivery and its site of action. At this point, it is important to discuss that conventional dosage forms can achieve local administration of the therapeutic compound. For example, when the desired active agent action is skin or mucosal, it is possible to administer the drug at the desired site using an ointment, lotion, cream, or gel form. Moreover, the direct injection of an anti-inflammatory agent into a joint is another example of site-specific delivery that is achievable without having recourse to a highly specialized drug delivery and targeting system (Hillery, Lloyd, & Swarbrick, 2001).

However, directing an active agent to a site of action via the route of administration does not ensure its retention at this site. When it is necessary to reach tissues sites (e.g., tumors) or internal organs exclusively (e.g., kidneys, heart, liver, and brain), the process is more complicated. In this case, the active agent needs to leave the systemic circulation and specially accumulate at targets inside the body. To accomplish this aim, physical means such as direct injection (e.g., intratumor injection) or implants (e.g., Gliadel, a polymer wafer that releases the drug when placed in the brain) are used (Siegel & Rathbone, 2012; Perrie & Rades, 2010; Aoki et al., 2014). These strategies do not ensure precise targeting and retention at the required site. It is very important to say that most active agents do not possess the required physicochemical characteristics to enter target cells. Thus, mechanisms to achieve this must be considered in the design of a targeting system if intracellular delivery is required (Perrie & Rades, 2010).

Drug targeting technology aims for precise drug targeting to the exact site of action. Recent developments in gene therapy, drug discovery, genomics and proteomics, as a consequence of the recent developments in molecular biology and chemistry, have provided opportunities for more effective site-specific drug targeting and bioresponsive drug delivery.

Direct administrations at or near the site of action, such those provided by therapeutic systems developed for eye and coronary artery delivery, direct injection of active agent into solid tumors or wound sites, timed release of growth and angiogenesis factors for vascularization of surgically implanted tissue engineered constructs, materials with bioresponsive features (temperature, pH, ionic strength, enzyme presence, etc.), nanocarriers with tumor affinity, ligands that bind to specific cell surface features, and antibodies raised against antigens expressed at the cell surface, are examples of this technology (Siegel & Rathbone, 2012).

4.8.3 Requirements

An effective drug targeting system should be non-toxic and therapeutically acceptable, and there must be no non-specific interactions with biological components, cells, tissues or organs. Therefore, the drug targeting system should be specific to the physiological target, retain the active agent during transit to the site, and release the active agent just at the target. In addition, the active agent must be able to be accessed and retained at the target site.

To accomplish these functions, the targeting systems can be soluble or particulate carrier systems. Soluble carriers are constituted of natural or synthetic hydrophilic polymers and antibodies, where the active agent is conjugated with the carrier. Particulate systems include liposomes, microspheres and nanoparticles, in which the active agent is either surface-bound or entrapped within the carrier.

These targeting systems have given an impression that they find target sites similar to a smart missile, which may actively chase its target from miles away (Bae & Park, 2011). This is certainly not the case. For oral delivery, systems are available to achieve site-specific delivery within the gastrointestinal tract (e.g., targeting the drug to the small intestine, colon, or gut lymphatics). Drug delivery systems available for targeted oral delivery include those that use enteric coatings, prodrugs, osmotic pumps, colloidal carriers, and hydrogels (Hillery et al., 2001).

Drug targeting to specific sites in the body requires different delivery systems depending on the drug delivery route selected. When the active agent is administered by parenteral route, the targeting technologies aim to deliver it to specific targets in the body and also to protect it from degradation and premature elimination. For this to be possible, necessary elements include efficient drug loading into the carrier, sufficient residence in the circulation to reach intended sites of the body, retention by specific characteristics within intended sites, and drug release at the intended site within a time that allows for effective function of the drug.

Depending on the physiology of the target site, the therapeutic system may naturally accumulate at the site, constituting passive targeting. However, the system may be actively targeted to a site using a target-specific recognition component.

With the aim of reaching internal organs, the active agent must leave the central circulation by crossing the endothelium. Normally, therapeutic systems are administered by parenteral route and are confronted with physical, chemical, and immunological barriers. The body's natural defense system is prepared to detect, prevent, and remove foreign substances, such as particulate drug delivery systems and drug conjugates, from entering further into the body via systemic circulation. For example, molecules within the blood can bind to particulates and mark them for recognition and destruction by phagocytic cells within the immune system.

The targeting delivery will become effective when the system is not cleared too quickly and provides a pharmacokinetic profile that will allow the active agent to interact with its physiological target. The circulation half-life of the system must be appropriate to allow the random encounters of the delivery system with such a target.

Drug carrier systems are generally too large to be eliminated by the kidneys, as clearance through kidney excretion is limited to sizes below 10 nm. However, there are various mechanisms to support the removal of particulates and large macromolecules from the body, such as clearance by a mononuclear phagocyte system and opsonization (Perrie & Rades, 2010). Therefore, it is necessary to camouflage the active agent and its carrier from the systemic environment until the active reaches its target. For example, to avoid the uptake of the mononuclear phagocyte system and rapid clearance from circulation, the delivery systems should be formulated to be below 100 nm in size and/or to have a modified surface to minimize opsonization (e.g., using polyethylene glycol polymer chains attached to another molecules for coating of the system). In addition, considering that nowadays the physiological target may be intracellular, additional rates of transport across the endothelial lining of the blood circulation and across cell membranes may need to be accounted for.

Therefore, it is possible to enumerate some points about the achievement of the targeting delivery. Drug targeting does not improve all drug therapies. An active agent that has the same site of efficacy and toxicity will not improve with this strategy.

Active agents displaying high total clearance are generally more appropriated, because the active agent not retained at the target site will be cleared rapidly.

4.8.4 Types of drug targeting

There are different forms of targeting: passive and active.

In passive targeting, the distribution of the system is dictated by local physiological conditions or by the mononuclear phagocyte system. Thus, carrier systems are taken up or trapped in organs such as the lung, liver, and spleen by virtue of their size and their surface properties such as size and charge. This type of targeting can be used to target hepatic cells and spleen macrophages, and this site-specific mechanism of clearance is a feature of the immune system. For example, passive targeting therapeutically can be used for the treatment of lysosomal enzyme deficiencies (e.g., Gaucher's disease, caused by a deficiency of the enzyme glucozylceramidase) or the treatment of macrophage intracellular microbial, viral or bacterial disease (e.g., visceral leishmaniasis).

In active targeting, the active element is the interaction of surface ligands with specific associated receptors to aid affinity with targets. These targeting groups are generally covalent interactions, rather than noncovalent, attached to the active agent or the surface of the carrier. The carrier system should be designed to avoid the passive targeting route, mainly the phagocytes of the mononuclear phagocyte system. There are various molecules/receptors that can be used to perform active targeting: folate receptors over-expressed in tumor tissues, antibody-antigen targeting, lectin-glycoprotein targeting, and physical targeting such as magnetically controlled drug delivery (Perrie & Rades, 2010).

Targeting can also be further subdivided according to the objectives or orders. First order implies targeting to organs, second order targeting to tissues within organs and third order targeting to cells within targets.

An even higher order of targeting would be to specific organelles in cells, including the nucleus in the case of gene delivery (Ruenraroengsak, Cook, & Florence, 2010). In these situations, the cellular uptake and intracellular routing of the active agent is dependent on endocytosis (phagocytosis, pinocytosis, and receptor-mediated endocytosis) and cytoplasmic delivery. When the active agent needs to exert its action within the cytoplasm of a cell, where its target enzymes are located, it must escape the endosomal vesicles. Several strategies have been investigated to facilitate and enhance the release from the endosomal compartment, such as the use of systems that can destabilize the endosomal membrane at low pH and inclusion of peptides or proteins in the system that can fuse with the endosomal membrane. Once in the citosol, the active agent can target the nucleus (normally when the DNA is the target and entering via the pores of the nuclear membrane) or the mitochondria (targeting can be achieved using mitochondrial localization proteins).

4.8.5 Drug targeting technology

Targets are not often directly in the immediate trajectory of the system. However, the system must release the therapeutic agent at the correct site and over an appropriate period to achieve an optimal effect. To reach this aim, the system can be subjected to the modification of size, surface nature, shape, and flexibility. There is the possibility of mimicking nature through its natural targeting systems, such as utilizing modified low-density lipoproteins and engineered viruses.

A specific interaction between the carrier system and target is necessary for specific uptake to occur. The specific interaction between the system and a receptor expressed by, or in the vicinity of, a target is a randomly determined process. The higher the strength of the bond between system and receptor, the greater the statistical chance for a useful interaction that leads to internalization.

The likelihood of such system–target interactions can be increased by the grafting of specific ligands to the system surface. Moreover, the forces of attraction between any system and its target come into play only at the nanometer range (Ruenraroengsak et al., 2010). According to all these considerations, the targeting technology includes the use of (Hillery et al., 2001; Bruschi & de Freitas, 2005; Bies, Lehr, & Woodley, 2004):

- Soluble carriers, such as monoclonal antibodies, dextrans, and soluble synthetic polymers
- Particulate carriers, such as liposomes, niosomes, micelles, micro- and nanoparticles, and dendrimers
- Target-specific recognition moieties, such as monoclonal antibodies, carbohydrates, and lectins
- Antibody-directed enzyme/prodrug therapy (ADEPT)
- Virus-directed prodrug/enzyme therapy (VDEPT)
- Chemical drug delivery systems

Sometimes, such as for cancer treatment, cell internalization for nanocarriers is essential for effective drug delivery, in addition to the enhanced permeation and retention (EPR) effect (Bae & Park, 2011; Fang, Nakamura, & Maeda, 2011). This effect was first reported by Matsumura and Maeda in 1986, and their investigations over the years showed that most solid tumors have blood vessels with defective architecture, producing extensive amounts of various vascular permeability factors. Enhanced vascular permeability will ensure a sufficient supply of nutrients and oxygen to tumor tissues for rapid growth. The EPR effect considers the unique anatomical–pathophysiological nature of tumor blood vessels that facilitates the transport of macromolecules (larger than 40 kDa) into tumor tissues (Matsumura & Maeda, 1986). This phenomenon does not occur in normal tissues (Maeda, Wu, Sawa, Matsumara, & Hori, 2000). Therefore, the EPR effect is considered a landmark principle in tumor-targeting chemotherapy and is becoming an increasingly promising paradigm for anticancer drug development (Fang et al., 2011; Lammers, Kiessling, Hennink, & Storm, 2012).

In the drug design process, the integration of considerations for targeting and drug delivery may allow the development of therapeutic systems that are potent and non-toxic, with the advantage that their chemical structure dictates the targeting of the active agent to a particular site of action through enzyme-based chemical delivery systems using prodrugs. Improved selective localization of anticancer agents to neoplastic tissue may be achieved using non-toxic prodrugs that release the active drug within the tumor as a result of enhanced enzyme activity in the cell.

Problems associated with targeted delivery systems are detected and require alterations in targeting strategies. Examples include the rapid clearance of targeted systems (especially antibody-targeted systems), immune reactions against intravenous administered systems, target-tissue heterogeneity, down regulation and sloughing of surface epitopes, diffusion and redistribution of released active agent leading to non-specific accumulation, and problems of insufficient localization of targeted systems into the tumor cells.

Thus, for the successful development of targeting drug delivery systems, it is necessary to include the development of suitable carriers and of drug entities that meet the required chemical and pharmacokinetic profile. The efforts need to aim for the resolution of the following fundamental issues (Petrak, 2005):

- The drug–carrier system must avoid non-specific interactions in the vascular compartment.
- The system should retain its ability to accumulate at the target site(s) and be in a form capable of acting on its pharmacological activity target.
- Active agents need to be selected, or rather designed, to have the pharmacokinetic properties compatible with the demands of target-selective drug delivery (especially drug retention at the site of delivery and its ability to access its site of molecular action).

Nowadays, the application of antibodies currently offers the most exciting prospect for site-specific drug delivery. However, it is important to understand increasingly the biological barriers to targeting. Another issue is to study if the specific size, charge, hydrophobicity, elasticity, shape and targeting moieties of systems will suffice for the quantitative attack on some targets (Ruenraroengsak et al., 2010).

Targeting is a very important mechanism of the spatial control of drug delivery, and makes the development of multifunctional drug delivery systems possible, showing a promising future. Drug targeting can contribute to the new generation of drug delivery technologies, such as redox-based drug delivery, antibody-directed enzyme prodrug therapy (ADEPT), gene-directed enzyme prodrug therapy (GDEPT), chronopharmacology, biosensors, stimuli-sensitive intelligent hydrogels, temperature-sensitive gels, pH-sensitive gels, bioresponsive drug delivery systems, glucose-sensing hydrogels, microchip delivery systems, and genetically engineered cell implants (Hillery et al., 2001).

References

Amkraut, A., Eckenhoff, J. B., & Nichols, K. (1990). Osmotic delivery of peptides and macromolecules. *Advanced Drug Delivery Reviews*, *4*, 255–276.

Aoki, T., et al. (2014). A multicenter phase I/II study of the BCNU implant (Gliadel® wafer) for Japanese patients with malignant gliomas. *Neurologia Medico-Chirurgica*, *54*, 290–301.

Bae, Y. H., & Park, K. (2011). Targeted drug delivery to tumors: Myths, reality and possibility. *Journal of Controlled Release*, *153*, 198–205.

Bies, C., Lehr, C.-M., & Woodley, J. F. (2004). Lectin-mediated drug targeting: History and applications. *Advanced Drug Delivery Reviews*, *56*, 425–435.

Bruschi, M. L., & de Freitas, O. (2005). Oral bioadhesive drug delivery systems. *Drug Development and Industrial Pharmacy*, *31*, 293–310.

Colombo, P. (1993). Swelling-controlled release in hydrogel matrices for oral route. *Advanced Drug Delivery Reviews*, *11*, 37–57.

Crank, J. (1975). *The mathematics of diffusion* (2nd ed.). Oxford: Claredon Press.

Ehrlich, P. (1954). The partial function of cells. *International Archives of Allergy and Applied Immunology*, *5*, 67–86.

Fan, L. T., & Singh, S. K. (1989). *Controlled release: Aquantitative treatment*. Berlin: Springer.

Fang, J., Nakamura, H., & Maeda, H. (2011). The EPR effect: Unique features of tumor blood vessels for drug delivery, factors involved, and limitations and augmentation of the effect. *Advanced Drug Delivery Reviews*, *63*, 136–151.

Florence, A. T., & Attwood, D. (2006). Physicochemical principles of pharmacy (4th ed.). London: Pharmaceutical Press.

Higuchi, T. (1961). Rate of release of medicaments from ointment bases containing drugs in suspensions. *Journal of Pharmaceutical Sciences*, *50*, 874–875.

Higuchi, W. I. (1962). Analysis of data on the medicament release from ointments. *Journal of Pharmaceutical Sciences*, *51*(8), 802–804.

Hillery, A. M., Lloyd, A. W., & Swarbrick, J. (2001). *Drug delivery and targeting for pharmacists and pharmaceutical scientists*. Oxford: Taylor & Francis.

Lammers, T., Kiessling, F., Hennink, W. E., & Storm, G. (2012). Drug targeting to tumors: Principles, pitfalls and (pre-) clinical progress. *Journal of Controlled Release*, *161*, 175–187.

Maeda, H., Wu, J., Sawa, T., Matsumara, Y., & Hori, K. (2000). Tumor vascular permeability and the EPR effect in macromolecular therapeutics: A review. *Journal of Controlled Release*, *65*, 271–284.

Mathiowitz, E. (1999). *Encyclopedia of controlled drug delivery*. New York: Wiley.

Matsumura, Y., & Maeda, H. (1986). A new concept for macromolecular therapeutics in cancer chemotherapy: Mechanism of tumoritropic accumulation of proteins and the antitumor agent SMANCS. *Cancer Research*, *46*, 6387–6392.

Mills, J. K., & Needham, D. (1999). Targeted drug delivery. *Expert Opinion on Therapeutic Patents*, *9*, 1499–1513.

Omidian, H., & Park, K. (2008). Swelling agents and devices in oral drug delivery. *Journal of Drug Delivery Science and Technology*, *18*(2), 83–93.

Ottenbrite, R. M., & Kim, S. W. (2002). *Polymeric drugs and drug delivery systems*. Boca Raton: CRC Press.

Peppas, N. A., & Narasimhan, B. (2014). Mathematical models in drug delivery: How modeling has shaped the way we design new drug delivery systems. *Journal of Controlled Release*, *190*, 75–81.

Perrie, Y., & Rades, T. (2010). *Pharmaceutics: Drug delivery and targeting*. London: Pharmaceutical Press.

Petrak, K. (2005). Essential properties of drug-targeting delivery systems. *Drug Discovery Today*, *10*(23/24), 1667–1673.

Ruenraroengsak, P., Cook, J. M., & Florence, A. T. (2010). Nanosystem drug targeting: Facing up to complex realities. *Journal of Controlled Release*, *141*, 265–276.

Siegel, R. A., & Rathbone, M. J. (2012). Overview of controlled release mechanisms. In J. Siepmann, R. A. Siegel, & M. J. Rathbone (Eds.), *Fundamentals and applications of controlled release drug delivery* (pp. 19–46). New York: Springer.

Theeuwes, F. (1975). Elementary osmotic pump. *Journal of Pharmaceutical Sciences*, *64*, 1987–1991.

Theeuwes, F., & Ayer, A.D. (1978), U.S. Pat. 4,077,407, to ALZA Corp.

Vyas, S. P., et al. (2001). Ligand-receptor-mediated drug delivery: An emerging paradigm in cellular drug targeting. *Critical Reviews in Therapeutic Drug Carrier Systems*, *18*(1), 1–76.

Wang, B., Siahaan, T., & Soltero, R. A. (2005). *Drug delivery: Principles and applications*. Hoboken: Wiley.

Wilson, C. G., & Crowley, P. J. (2011). *Controlled release in oral drug delivery*. London: Springer.

Mathematical models of drug release

5

5.1 Introduction

Over the past three decades, there have been significant advances in the area of drug delivery with the development of controlled release dosage forms.

Drug release is an important property of a therapeutic system, constituting a prerequisite to absorption of the therapeutic agent and one that contributes to the rate and extent of active availability to the body. The release patterns comprise the zero- and first-order rate. In addition, there are those that provide an initial rapid dose, followed by zero- or first-order release of the sustained component.

The aim of these systems is to maintain the concentration of the therapeutic agent in the blood or in target tissues at a desired value as long as possible, exerting a control on the drug release rate and duration. Therefore, the controlled release system can initially release a fraction of the dose contained in order to attain rapidly the effective therapeutic concentration of the drug (burst release). After, a well-defined behavior of drug release kinetics can supply the maintenance of effective drug concentration level.

Therefore, controlled release dosage forms enable pharmacists and engineers to work together with the aim of designing controlled drug delivery systems. To provide particular, predetermined release profiles, it is necessary to know the exact mass transport mechanisms involved in drug release, and to predict quantitatively the resulting drug release kinetics.

Many times, it is possible to acquire a mathematical equation that describes the dependence of release in function of time. The use of this tool is very useful to predict the release kinetics before the release systems are realized. This analytical solution conduces to many models that have been used to design a number of simple and complex drug delivery systems and devices and to predict the overall release behavior.

Mathematical models are an important tool to design pharmaceutical formulations, evaluate drug release processes in vitro and in vivo and, in general, come up with the optimal design for new systems (Peppas & Narasimhan, 2014). They allow the measurement of some important physical parameters (e.g., drug diffusion coefficient) and resort to model fitting on experimental release data. The amount and type of active agent, polymer and adjuvants as well as the size and shape of the system designed to achieve a certain drug release profile can be predicted theoretically (Siepmann & Peppas, 2012).

It is very important to know how to use these equations to understand the different factors that affect the dissolution velocity and how the dissolution behaviors can vary and influence the efficiency or the therapeutic regimen of patients. Mathematical

Strategies to Modify the Drug Release from Pharmaceutical Systems. http://dx.doi.org/10.1016/B978-0-08-100092-2.00005-9

equations enable the quantitative interpretation of the values obtained from a dissolution or drug release assay. The equation needs to have the ability to transform the release curve in function of some other parameter related to the dosage form under analysis. Therefore, the models have been used mainly to predict the release of the encapsulated molecule(s) in function of time. Therefore, mathematical evaluation of drug release kinetics adds value, ensuring optimal design of pharmaceutical formulation(s) as well as understanding release mechanism(s) through experimental verification (Peppas & Narasimhan, 2014).

The precise combination of experimental observations with models that make it possible to acquire the underlying physics will provide the understanding about the release mechanism. When the analytical solutions are explicit, it is possible to relate the dependent and independent variables. However, this dependence is not obvious in the case of implicit analytical solutions. The understanding of the effect of certain independent variables on particular dependent variables is much easier, compared to numerical solutions. Physically more realistic models are mathematically more complex, and very often it is difficult to find analytical solutions of the respective set of equations (Narasimhan & Peppas, 1997).

At this point, it is clear that the development of mathematical modeling requires the comprehension of all the phenomena affecting drug release kinetics. The model can be simply considered a mathematical metaphor of some aspects of reality that, in this case, identifies with the ensemble of phenomena ruling release kinetics.

According to what was previously described, the drug release from a modified release system can be controlled by various methods, such as dissolution, diffusion, partitioning, osmosis, swelling, and erosion. Diffusion of the active agent is a strong function of the structure through which the diffusion takes place, and diffusion equations can be predicted by many existing models. For example, it is important to consider the polymer morphology (Langer & Peppas, 1983).

Generally, when a hydrophilic drug is incorporated in a matrix, the release occurs easily by diffusion, compared to another hydrophobic or less water-soluble drug. The hydrophobic drug release normally is associated with swelling and/or matrix erosion. Differences in solubility, in the factors that determine the solubility and the participation of different processes explain why the same active agent, when incorporated in a polymeric matrix, can present different behaviors (Agnes & Gonzales Ortega, 2003).

Considering that qualitative and quantitative changes in a system may alter drug release and *in vivo* performance, the development of tools that facilitate product development by reducing the necessity of biostudies is always desirable. In this regard, the use of *in vitro* drug dissolution data to predict *in vivo* bioperformance can be considered the rational development of controlled release formulations (Dash, Murthy, Nath, & Chowdhury, 2010).

Statistical methods can be used to investigate the kinetics of drug release from controlled release formulation, using the exploratory data analysis method, repeated measures design and multivariate approach (e.g., multivariate analysis of variance (MANOVA)) (Mauger, Chilko, & Howard, 1986; Polli, Rekhi, Augsburger, & Shah, 1997). The exploratory data analysis method is the first step to compare dissolution profile data in both graphical and numerical form, constituting a useful manner to obtain improved understanding of the dissolution data of the controlled release system. Briefly, the release profile data are illustrated graphically by plotting the mean release profile data for each preparation, with error bars extending to two standard errors at each dissolution time point. The data of the dissolution profiles are summarized numerically, and 95% confidence intervals for the differences in the mean dissolution profiles at each dissolution time point are evaluated (O'Hara, Dunne, Butler, & Devane, 1998).

The complexity of the involved process justifies the frequent utilization of empirical equations. However, statistical methods are recommended when the aim is to evaluate the similarities between dissolution/release behaviors (Agnes & Gonzales Ortega, 2003). Methods independent of model (e.g., difference factor (f_1), similarity factor (f_2)) are also utilized (Costa, 2001).

However, model-dependent methods, based on different mathematical functions, are very useful to describe the release profile. Once a suitable function has been selected, the dissolution profiles are evaluated depending on the derived model parameters. Between the main-release kinetic models are zero order, first order, Higuchi, Hixson–Crowell, Korsmeyer–Peppas, Baker–Lonsdale, Weibull, Hopfenberg, and Gompertz.

5.2 Zero-order kinetics

Excursions of drug concentration in the blood, particularly when the active agent is rapidly absorbed and rapidly eliminated, may lead to periods of underexposure or overexposure. When an active agent is administered from rapid-release dosage forms, a little time is spent inside the so-called therapeutic range. Therefore, frequent repetitive dosing is necessary, with the aim of the maintenance of the effective drug concentration, and compliance and control are difficult.

Therapeutic systems may release the active agent at a constant rate for a certain period of time, and are desirable in order to minimize these variations. Drug delivery technology has made possible the equalized administration and elimination of an active, resulting in a uniformity of drug concentration in plasma or tissue over the predetermined period of time.

Considering that dissolution is a kinetic process, the velocity of dissolution reflects the amount of drug dissolved during this time. The velocity in which a solid dissolves in a solvent can be described as

$$\frac{\mathrm{d}M}{\mathrm{d}t} = \frac{DS}{l}\left(C_{\mathrm{s}} - C\right) \tag{5.1}$$

or

$$\frac{\mathrm{d}C}{\mathrm{d}t} = \frac{DS}{Vl}\left(C_{\mathrm{s}} - C\right) \tag{5.2}$$

where M is the mass of solute dissolved during the time t, $\mathrm{d}M/\mathrm{d}t$ is the velocity of mass dissolved (mass/time), D is the coefficient of diffusion of solute in the solution, S is the solute area exposed, l is the thickness of the diffusion layer, C_s is the solid solubility, and C is the solute concentration into the solution on time t. Therefore, dc/dt is the velocity of dissolution and V is the volume of solution.

According to the theory of mass transport, there is an aqueous diffusion layer or a static liquid film of thickness on the surface of a solid that is dissolving, as described in Figure 4.1.

The dissolution of active agents contained in non-disintegrating dosage forms, considering a very slow drug release (with no changes in the equilibrium conditions), can be represented by the equation

$$W_0 - W_i = K \cdot t \tag{5.3}$$

where W_0 is the initial mass of active agent in the dosage form; W_i is the remaining mass in the dosage form, on time t and K is the constant of proportionality.

Dividing this equation by W_0 and simplifying, it is possible to infer that

$$f_i = K_0 t \tag{5.4}$$

where $f_i = 1 - (W_i/W_0)$ represents the fraction of active agent dissolved during the time t and K_0 is a constant of apparent velocity of dissolution.

The graphical representation of Equation (5.4) results in a straight, in which the angular coefficient corresponds to K_0 (Figure 5.1).

This relationship can be found in various drug delivery systems (e.g., transdermal slow release matrix, coated, and osmotic systems). The systems that display this type of release behavior are suitable for prolonged drug release (Figure 5.2).

Moreover, Equation (5.4) can be described in terms of concentration

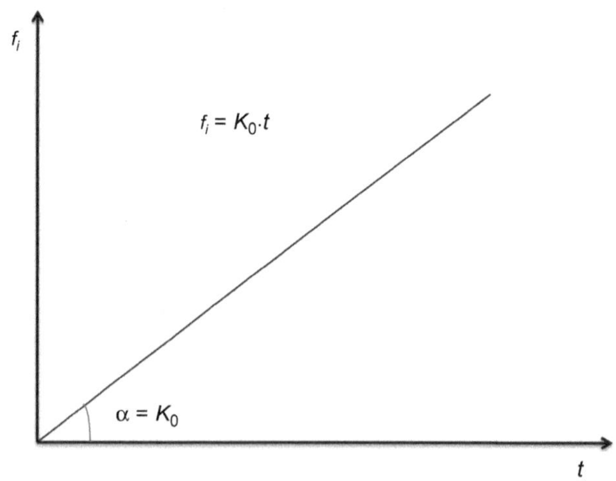

Figure 5.1 Graphical representation of zero-order kinetics: f_i is the fraction of the drug dissolved during the time t and K_0 is the angular coefficient or the apparent velocity of dissolution.

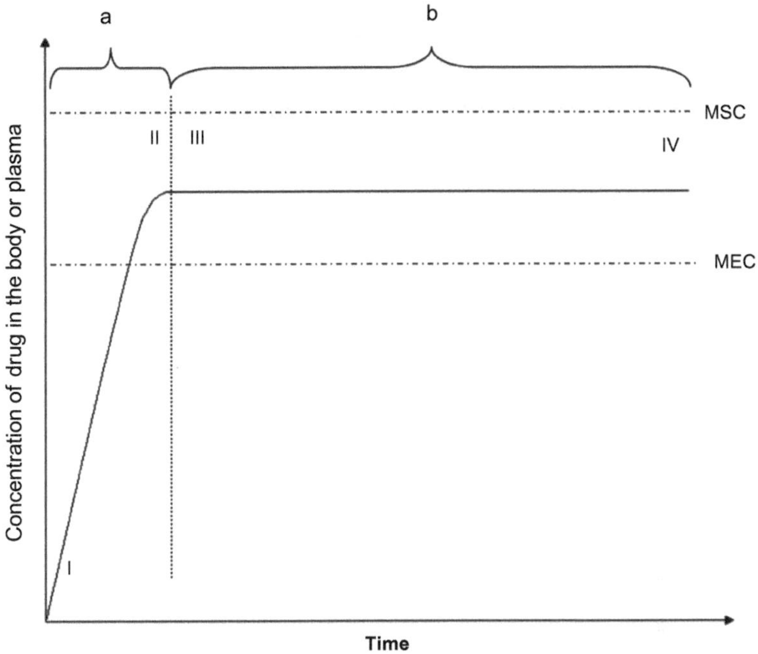

Figure 5.2 Drug level versus time profile of zero-order, controlled release: (a) absorption phase; (b) elimination phase; (I and II) rate of drug absorption higher than rate of drug elimination; (III and IV) rate of drug input equal to rate of drug output; (MSC) maximum safe concentration; (MEC) minimum effective concentration.

$$C_t = C_0 + K_0 t \tag{5.5}$$

where C_t represents the amount of active agent released during the time t, C_0 is the initial concentration of active released (generally, $C_0=0$), and K_0 is the zero-order constant.

For zero-order kinetics, the release of an active agent is only a function of time and the process takes place at a constant rate independent of active agent concentration.

Moreover, zero-order oral drug release can be achieved by oral route by coating a core tablet. When the coating is permeable to both active agent and water, after swallowing, the core becomes hydrated and the active dissolves until it reaches its saturation concentration or solubility. The active is located in a saturated reservoir and is released by partition from the reservoir into the membrane, diffusing through the membrane, and reaching the gastrointestinal fluid. The constant release rate is due to the saturation in the core, producing a stationary concentration gradient across the membrane. When this concentration falls below saturation, the drug release rate decays to zero.

The osmosis property can also be used to provide zero-order release and has been an alternative to dissolution, partition, and diffusion. Tablets or capsules consisting of a core of active agent surrounded by a membrane, which is permeable to water but not to the active, have been proposed. The systems possess a small hole that is drilled into the membrane. Upon administration, water of fluids is osmotically imbibed into the core through the semipermeable membrane, dissolving the active. Therefore, a constant osmotic pressure gradient is established between the core and the external medium. The water influx does the active release through the hole at a constant rate. Eventually, drug concentration falls below its solubility, and the rate of osmotic pumping decays.

If the device is formulated as a slab, then release will be approximately zero order, because each time interval will correspond to the erosion of a layer of polymer and release of drug incorporated in that layer. The erosion rate of cylinders and spheres decreases with time, however, due to reduction in exposed surface area. In principle, drug release correlates with erosion.

Transdermal systems can be developed to release the active agent at zero-order kinetics. A drug reservoir can be constituted by a variety of materials, ranging from mineral oil to complex formulations such as aqueous/alcoholic gels. A reservoir system should provide zero-order release of the drug over the entire delivery period, which requires that the reservoir material remain saturated with the drug over the period of product application. This can usually be achieved by formulating the drug as a suspension (Walters & Brain, 2009).

5.3 First-order kinetics

This model has been used to describe absorption and/or elimination of a variety of therapeutic agents. However, it is difficult to define first-order kinetics using a basic theory. In this sense, first-order release kinetics states that change in concentration with respect to change on time is dependent only on concentration

$$\frac{\mathrm{d}C}{\mathrm{d}t} = -KC \tag{5.6}$$

where C is the concentration of drug in the drug molecule and k is the first-order release constant.

The phenomenon of dissolution of a solid particle in a liquid implies a superficial action and can be described by the (Noyes & Whitney, 1897)

$$\frac{\mathrm{d}c}{\mathrm{d}t} = K\left(C_{\mathrm{s}} - C\right) \tag{5.7}$$

where C is the concentration of solute at the time t, C_{s} is the equilibrium solubility at the temperature of the process, and K is the first-order constant.

Equation (5.7) can be modified, adding the contact or access surface variable

$$\frac{\mathrm{d}c}{\mathrm{d}t} = K_1 S\left(C_{\mathrm{s}} - C\right) \tag{5.8}$$

where K_1 is a new constant of proportionality.

Considering Fick's first law, it is possible to get a relationship for the constant K_1:

$$K_1 = \frac{D}{Vh} \tag{5.9}$$

where D is the diffusion coefficient in the release medium, V is the volume of the liquid of the release medium, and h the thickness of the diffusion layer.

Hixson and Crowell proposed a modification of Noyes–Whitney (Equation (5.7)):

$$\frac{\mathrm{d}W}{\mathrm{d}t} = KS\left(C_\mathrm{s} - C\right) \tag{5.10}$$

where W is the amount of drug soluble on time t, $\mathrm{d}W/\mathrm{d}t$ is the ratio of the solute passing to the solution on time t, S is the surface of dissolution, and K the constant of proportionality.

Multiplying both terms of the equation by V, considering that K equals K_1V and $k = D/h$, the equation proposed by Hixson and Crowell can be described as

$$\frac{\mathrm{d}W}{\mathrm{d}t} = \frac{KS}{V}\left(VC_\mathrm{s} - W\right) = k\left(VC_\mathrm{s} - W\right) \tag{5.11}$$

where $k = k_1S$.

When a therapeutic system maintains a constant area under sink conditions, it is possible to use the last equation (Equation (5.10)) and, after integration, it results as

$$W = VC_\mathrm{s}\left(1 - \mathrm{e}^{-kt}\right) \tag{5.12}$$

This equation can be transformed by applying decimal logarithms in both terms:

$$\log\left(VC_\mathrm{s} - W\right) = \log VC_\mathrm{s} - \frac{kt}{2.303} \tag{5.13}$$

However, the term VC_s is equal to the mass Q, and Equation (5.13) can be described as

$$\log Q_1 = \log Q_0 + \frac{k_1 t}{2.303} \tag{5.14}$$

where Q_1 is the amount of active agent released on time t, Q_0 is the initial amount of drug dissolved, and K_1 is the first-order constant.

Equation (5.14) corresponds to a linear function, and the graph of Napierian or decimal logarithm of the mass released of the drug will result in a straight, with angular coefficient $K_1/2.303$ and linear coefficient equal to $\log Q_0$ (Figure 5.3).

There are a variety of therapeutic systems that display this type of first-order release. For soluble active agents incorporated in a porous matrix, the amount of drug released is proportional to the amount of remaining drug in the matrix. Thus, the amount of active released tends to decrease in function of time.

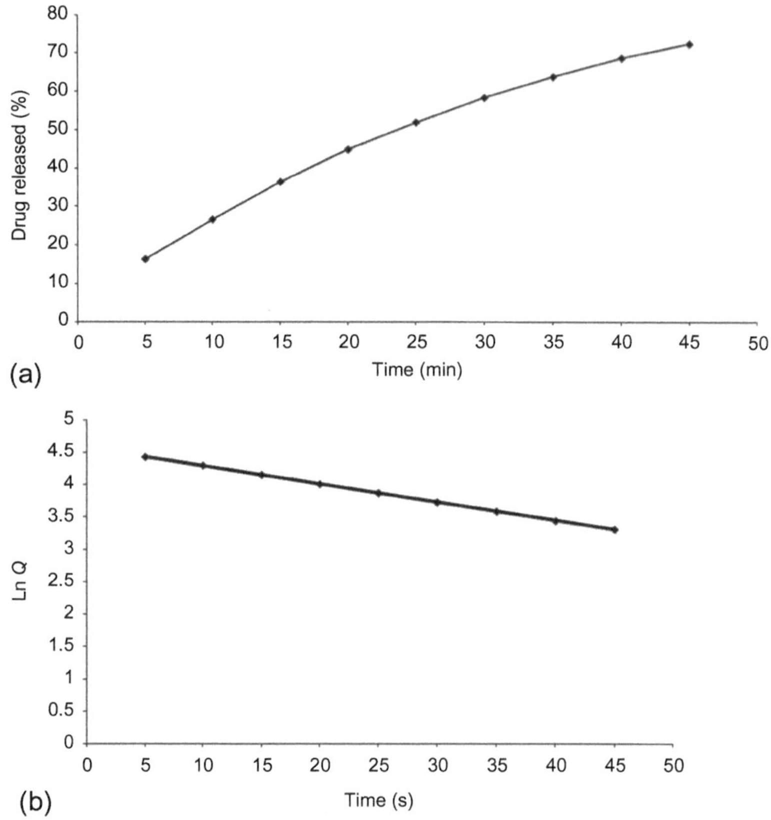

Figure 5.3 Graphical representation of first-order model of drug release: (a) cumulative drug released (%) versus time (s); (b) Napierian logarithm of the non-released fraction of drug (Ln *Q*) versus time (s), showing good linearity.

5.4 Higuchi model

A mathematical model to describe the drug dissolution from matrix systems was not developed until the 1960s. From 1961 to 1963, there was a great advance in the development of mathematical models to understand drug release.

In 1961, Higuchi published probably the most famous and most often used mathematical equation to describe the release rate of drugs from matrix systems. The rate of drug release from ointment bases (planar systems) containing drugs in suspension was the object of investigation (Higuchi, 1961). In 1962, the analysis of data on the drug release from ointments was accomplished (Higuchi, 1962). Moreover, in 1963, the "mechanism of sustained-action medication" was proposed (Higuchi, 1963).

These three works enabled the development of many theoretical models to describe the release of active agents that are less soluble as well as very soluble, contained in solid and semi-solid matrices.

Mathematical relationships developed by Higuchi are related to particles of active dispersed in homogeneous matrices submitted to a diffusing medium. Considering the dissolution of a lipophilic, homogeneous, planar matrix, the relationship initially proposed by Higuchi is

$$f_1 = Q = \sqrt{D(2C - C_s)C_s t} \tag{5.15}$$

where Q is the amount of drug released on time t by area unit, C is the initial amount of drug contained in dosage form, C_s is the solubility of active agent in the matrix medium and D is the diffusion coefficient in the matrix medium.

This relationship is valid over the time of dissolution, except when the drug-released levels tend to saturate the liquid medium contained in the matrix.

Despite the conditions adopted (lipophilic, homogeneity and planar form), this model showed itself transposable for other types of dosage forms. Considering the homogeneous and spherical, heterogeneous and planar, and heterogeneous and spherical systems, Higuchi proposed other models considering a low concentration of active agent in the matrix, where the solubility and release happen through the porosity of the matrix:

$$f_1 = Q = \sqrt{\frac{D\varepsilon}{\tau}(2C - \varepsilon C_s)C_s t} \tag{5.16}$$

where Q is the amount of drug released on time t by area unit, ε is the porosity of matrix, τ is the capillary tortuosity factor, C is the initial amount of drug contained in the dosage form, C_s is the solubility of active agent in the matrix medium, and D is the diffusion coefficient in the matrix medium.

This model assumes no important alteration of matrix structure during its contact with water.

Higuchi's work of 1962 proposed an equation applied to a matrix saturated with a drug, where C_0 is the concentration of diffusing liquid contained in a porous matrix:

$$f_1 = Q = \sqrt{2C_0\varepsilon\frac{Dt}{\tau\pi}} \tag{5.17}$$

Therefore, it is obvious to conclude that the amount of drug released is proportional to the square root of time:

$$f_1 = Q = K_H\sqrt{t} \tag{5.18}$$

where K_H is the release constant of Higuchi.

Equation (5.18) is the simplified Higuchi model, which relates the concentration of active agent to the square root of time, representing a linear function (Figure 5.4).

There are assumptions that need to be followed with the use of the Higuchi model (Omidian & Park, 2010; Siepmann & Peppas, 2012):

(1) The matrix contains an initial drug concentration much higher than the solubility of the drug.
(2) The diffusion is unidirectional, because the edge effects are negligible.
(3) The thickness of the dosage form is much larger than the size of the drug molecules.

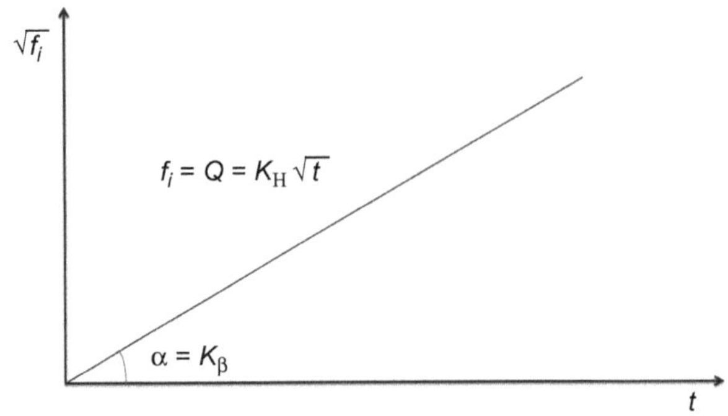

Figure 5.4 Square root of fraction of active agent released $\left(\sqrt{f_i}\right)$ versus the time t.

(4) The swelling or dissolution of the matrix is negligible.
(5) The diffusivity of the drug is constant.
(6) The perfect sink conditions are attained in the release environment.

The first assumption in which the initial drug concentration in the system is much higher than the solubility of the drug provides the basis for the justification of the applied pseudo-steady-state approach (Siepmann & Peppas, 2012).

This mathematical treatment considering the drug diffusivity constant is only valid in the case of polymers that do not significantly swell upon contact with water (e.g., ethylcellulose).

When the total dissolution of the drug molecule in solution yields a resulting concentration that is much lower than that of saturation (typically from 1/10 to 1/3 of saturation), the dissolution system is working under sink conditions, and the driving force of dissolution is higher.

Higuchi models enabled the development of a variety of mathematical approaches, which were important to classify the possible release profile of active agents in dosage forms. The Korsmeyer and Peppas model and the modifications proposed by Masaro and Zhu are examples (Peppas & Narasimhan, 2014).

5.5 Hixson–Crowell model

Hixson and Crowell (1931) discovered that a group of particles' regular area is proportional to the cube root of its volume. Using this relationship, they proposed an equation:

$$\sqrt[3]{W_0} = \sqrt[3]{W_i} + K_{HC}t \tag{5.19}$$

where W_0 is the initial amount of the drug in the system; W_i is the amount remaining in the system on time t; and K_{HC} is the constant of incorporation, which relates surface and volume.

This equation applies to dosage forms such as tablets, in which the dissolution happens in planes parallel to the surface of the dosage form; this surface decreases proportionally over time and the geometrical form maintains constant.

Considering the concentration, drug density and the number of spherical particles, Equation (5.14) can also been described as

$$\sqrt[3]{W_0} = \sqrt[3]{W_i} + \frac{K'\sqrt[3]{N}DC_s t}{\delta} \tag{5.20}$$

where N is the number of particles; K' is a constant that relates surface, form and particle density; D is the diffusion coefficient; C_s is the solubility at the equilibrium (saturation concentration) at the temperature of process; and δ is the thickness of the diffusion layer.

The surface factors of cubic or spherical particles are constant if the dissolution is constant over the entire system.

When the particles are irregular, this behavior is different and a deviation in relationship to values theoretically provided through Equation (5.19) appears. Moreover, dividing by $\sqrt[3]{W_0}$, it is possible to simplify

$$\sqrt[3]{1-f_i} = 1 - K_\beta t \tag{5.21}$$

where $f_1 = 1 - (W_i/W_0)$ and represents the fraction of drug dissolved on time t and K_β is a release constant.

Equation (5.21) represents a linear function when the cubic root of the non-released fraction is related with the time. This is true when the equilibrium conditions are not modified and the surface of the particle decreases proportionally over time (Figure 5.5).

When this model is used, it is assumed that the drug release is limited by dissolution velocity and not by diffusion, which can occur through the polymeric matrix. Thus, the Hixson–Crowell equation applies to pharmaceutical dosage forms such as

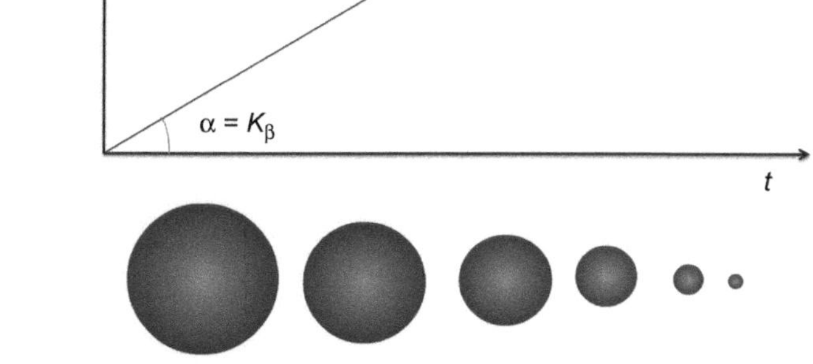

Figure 5.5 Cubic root of fraction of active agent not released $\left(\sqrt[3]{1-f_i}\right)$ versus the time t and the illustration of proportional decreasing of surface particle.

tablets, considering that dissolution occurs in planes parallel to the surface of the active agent if the tablet dimensions decrease proportionally, but with maintenance of the geometrical characteristics.

5.6 Ritger–Peppas and Korsmeyer–Peppas model (Power law)

Power law is a more comprehensive semi-empirical equation to describe drug release from polymeric systems. This model was developed by Korsmeyer, Gurny, Doelker, Buri, and Peppas (1983) and Ritger and Peppas (1987) as a semi-empirical model, establishing the exponential relationship between the release and the time:

$$f_1 = \frac{M_i}{M_\infty} = Kt^n \tag{5.22}$$

where f_1 is the amount of drug released, M_∞ is the amount of drug at the equilibrium state (sometimes very close to the amount of drug contained in the dosage form at the beginning of the release process), M_i is the amount of drug released over time t, K is the constant of incorporation of structural modifications and geometrical characteristics of the system (also considered the release velocity constant), and n is the exponent of release (related to the drug release mechanism) in function of time t.

This model was developed specifically for the release of a drug molecule from a polymeric matrix, such as a hydrogel. Over the last 25 years, many scientists have employed a modified form of this equation that contains the latency time (l), which marks the beginning of drug release from the system (El-Arini & Leuenberger, 1998; Ford et al., 1991; Kim & Fassihi, 1997; Pillay & Fassihi, 1999):

$$\frac{M_{(i-l)}}{M_\infty} = K\left(t-l\right)^n \tag{5.23}$$

or the logarithmical version:

$$\log\left(\frac{M_{(i-l)}}{M_\infty}\right) = \log K + n\log\left(t-l\right) \tag{5.24}$$

where M_∞ is the amount of drug at the equilibrium state, M_i is the amount of drug released over time t, K is the constant of incorporation of structural modifications and geometrical characteristics of the system (also considered the release velocity constant), n is the exponent of release (related to the drug release mechanism) in function of time t and l is the latency time.

In addition, when the drug release process is characterized by an abrupt increase of initial drug release (burst effect), the following equation was proposed (Kim & Fassihi, 1997):

$$\frac{M_i}{M_\infty} = Kt^n + b \tag{5.25}$$

where b is the burst effect.

The power law model is useful for the study of drug release from polymeric systems when the release mechanism is not known or when more than one type of phenomenon of drug release is involved (Peppas, 1985; Peppas & Narasimhan, 2014). Actually, it can be seen as a generalization of the observation of the superposition of two apparently independent mechanisms of drug transport, relaxation, and diffusion.

Depending on the value of n that better adjusts to the release profile of an active agent in a matrix system, it is possible to establish a classification, according to the type of observed behavior:

- Fickian model (Case I)
- Non-Fickian models (Case II, Anomalous Case and Super Case II)

In the Fickian model (Case I), $n = 0.5$ and the drug release are governed by diffusion. The solvent transport rate or diffusion is much greater than the process of polymeric chain relaxation. Equilibrium of absorption in the surface exposed of the polymeric system takes place rapidly, leading to conditions of time-dependent links. The kinetics of this phenomenon are characterized by diffusivity.

When $n = 1$, the model is non-Fickian (Case II), the drug release rate corresponds to zero-order release kinetics and the mechanism driving the drug release is the swelling or relaxation of polymeric chains. This is another extreme type of behavior. The diffusion of solvent through the system (jellified region) is very fast compared to the process of relaxation on the gel-vitreous polymeric interface, which marks the inner limit of penetrating or solvent dispersion at a determined time. The vitreous region has almost no penetrating solvent and the diffusivity of the solvent in the gelled region is higher. The solvent concentration profile (level) in the system has a distinct format: plane and zero in the vitreous nucleus, plane and equal to the level in the equilibrium with the external medium in the gelled region. At the end of Case II transport, a fast increase of absorption rate of solvent may sometimes be observed. In this situation, the transport Case II evolved to transport Super Case II, due to the expansion of forces exercised by swollen gel in the vitreous nucleus.

Moreover, when $0.5 < n < 1$, the model is non-Fickian or anomalous transport, and the mechanism of drug release is governed by diffusion and swelling. The diffusion and swelling rates are comparable. The rearrangement of polymeric chains occurring slowly and the diffusion process simultaneously cause the time-dependent anomalous effects.

Finally, the Super Case II model is characterized when $n > 1$, constituting an extreme form of transport. During the sorption process, tension and breaking of the polymer (solvent crazing) occurs. The outer layer of gel limits the vitreous nucleus, which in a cylindrical sample prevents the axial swelling of the gel, producing tension of compression on the nucleus. As the polymeric interface gel-vitreous moves to the nucleus, the tension increases until the nucleus breaks.

Table 5.1 summarizes this classification that constitutes very important information, because the different values of n are not always taken into account, leading to misinterpretations of experimental results.

To determine the exponent n, it is recommended to use the portion of release curve until the point where $\frac{M_i}{M_\infty} < 0.60$. In all cases, it is assumed that the release occurs

Table 5.1 **Interpretation of release models of active agent from polymeric matrices with different geometries (Kosmidis, Argyrakis, & Macheras, 2003; Peppas, 1985)**

Release mechanism model	Geometry	Release exponent (n)	Time in function of n
Fickian diffusion	Planar (thin films)	0.50	$t^{0.50}$
	Cylinders	0.45	$t^{0.45}$
	Spheres	0.43	$t^{0.43}$
Anomalous transport	Planar (thin films)	$0.50 < n < 1.0$	$t^{0.50<n<1.0}$
	Cylinders	$0.45 < n < 0.89$	$t^{0.45<n<0.89}$
	Spheres	$0.43 < n < 0.85$	$t^{0.43<n<0.85}$
Case I transport	Planar (thin films)	1.0	t[a]
	Cylinders	0.89	$t^{0.89}$
	Spheres	0.85	$t^{0.85}$
Super Case II transport	Planar (thin films)	$n > 1$	$t^{n>1}$
	Cylinders	$n > 0.89$	$t^{n>0.89}$
	Spheres	$n > 0.85$	$t^{n>0.85}$

[a] Zero-order release.

in just one direction and that the relationship between the width and thickness of the matrix is a maximum of 1:10 (Peppas, 1985).

Fickian diffusion generally occurs in polymeric matrices in which the value of the vitreous transition temperature of polymer (T_g) is less than the temperature of the environment. When the polymer is in the rubbery state, polymeric chains have high mobility, enabling easy solvent penetration.

However, Fickian diffusion is characterized by high velocity of solvent diffusion to the interior of the matrix and low velocity of polymeric relaxation. This behavior induces the formation of a gradient of solvent penetration. The solvent concentration decreases exponentially from the region totally swollen to the center of the matrix. The diffusing distance is proportional to the square root of time. It is possible to observe the Fickian diffusion in polymers with T_g higher than the ambient temperature, when a plasticizing is added to the system.

Non-Fickian processes are normally found in vitreous polymers, when the environmental temperature is less than the T_g. Between the three classes of non-Fickian diffusion (Case II, Anomalous Case and Super Case II transport), the main difference is the velocity of solvent diffusion. In Case II, the velocity of solvent diffusion is less than the polymeric relaxation process. In Anomalous transport, the velocity of solvent diffusion and the polymeric relaxation possess similar magnitudes. However, in Super Case II, the velocity of solvent diffusion is much higher, causing an acceleration of solvent penetration (Klech & Simonelli, 1989).

Transport Case II is observed when the solvent has a high affinity by matrix. The main characteristics of this release model are

(1) A fast increase in the concentration of solvent, with a rapid penetration to the center of the matrix

(2) A solvent concentration that remains constant both in the swollen portion and in the swelling border
(3) A solvent penetration border that is fast and advances with constant velocity, and a diffusion distance that is proportional to time
(4) An induction time in the Fickian concentration profile before the penetration of solvent to the center of the matrix, which is in a vitreous state

Klech and Simonelli (1989) showed that in the same system, the simultaneous effect of more than one transport mechanism may occur. The swelling of a polymeric matrix with spherical form with radius r displays a typical profile, where first Fickian diffusion takes place, then Case II and last, Super Case II. The borders of swelling of the spherical matrix were described as responsible for this behavior.

5.7 Brazel and Peppas model

In release systems controlled by swelling, the water absorption and drug release can be described by two-dimensional parameters using the diffusing number Deborah (De), with utility in many polymer/solvent systems.

The number De expresses the relationship between the characteristics of polymeric relaxation time and the characteristics of diffusion time, enabling the evaluation of the motility of water in relation to polymer relaxation, and the interfacial swelling index (Sw) that measures the water penetration in the polymeric chain in relation to the drug dispersed in the polymer:

$$\mathrm{De} = \frac{\lambda}{\theta} = \frac{\lambda D_{1,2}}{\left(\delta(t)\right)^2} = \frac{\text{Relaxation time}}{\text{Time of solvent diffusion}} \tag{5.26}$$

where λ is the relaxation time of the polymer and θ is the diffusion time of water characteristic for a swellable matrix, and is equivalent to diffusing distance divided by diffusing coefficient of water in the polymer $\left(\frac{\delta^2}{D_{1,2}}\right)$.

The water motility and Sw are sensitive to the swelling of the polymer and the drug properties.

When the swelling process is determined, by relaxation time (De >> 1) or by diffusion of water (De << 1), the following process is a function of time, characterizing a Fickian process. In contrast, when De is near 1, it is considered that the two processes are occurring simultaneously (Anomalous transport).

Sw is important to describe the equilibrium between the water entrance and the drug release:

$$\mathrm{Sw} = \frac{\upsilon \delta_{\mathrm{r}}}{D_{3,21}} = \frac{\text{Solvent motility}}{\text{Drug diffusion}} \tag{5.27}$$

where υ is the velocity of movement of the vitreous/rubbery border of the polymer, δ is the thickness of the swollen gel, and $D_{3,21}$ is the diffusion coefficient of the drug in the polymer. Sw is sensitive to the swelling of the polymeric structure, affecting water absorption and the drug property through the diffusion coefficient of the drug in the polymer. When Sw is significantly greater or less than 1, the water penetration or diffusion controls the release, and the time dependence will be according to Fick. However, if the value of Sw is near 1, the release is guided by Anomalous transport.

5.8 Baker and Lonsdale model

Baker and Lonsdale (1974) developed a mathematical model for drug release based on the Higuchi model. The model is described for controlled drug release from spherical matrices

$$\frac{3}{2}\left[1-\left(1-\frac{M_t}{M_\infty}\right)^{2/3}\right]-\frac{M_t}{M_\infty}=\frac{3D_{\mathrm{m}}C_{\mathrm{ms}}}{r_0^{\,2}C_0}-t \tag{5.28}$$

where M_t is the amount of the drug released at the time t, M_∞ is the amount released at the infinite time, D_{m} is the diffusion coefficient, C_{ms} the solubility of the drug in the matrix, r_0 is the radius of the spherical matrix, and C_0 is the initial concentration of the drug in the matrix.

For non-homogeneous matrices with fissures or capillaries capable of increasing the drug release, a modification of the previous equation was proposed (Seki et al., 1980):

$$\frac{3}{2}\left[1-\left(1-\frac{M_t}{M_\infty}\right)^{2/3}\right]-\frac{M_t}{M_\infty}=\frac{3D_{\mathrm{f}}C_{\mathrm{fs}}\varepsilon}{r_0^{\,2}C_0\tau}-t \tag{5.29}$$

where D_{f} is the diffusion coefficient, C_{fs} is the solubility of the drug in the liquid receptor, τ is the tortuosity factor of the capillary system, and ε is the matrix porosity.

The matrix porosity can be described as

$$\varepsilon=\varepsilon_0+KC_0 \tag{5.30}$$

where ε_0 is the initial porosity and K the constant of the specific volume of the system.

When ε_0 is little, the equation is

$$\frac{3}{2}\left[1-\left(1-\frac{M_t}{M_\infty}\right)^{2/3}\right]-\frac{M_t}{M_\infty}=\frac{3D_{\mathrm{f}}KC_{\mathrm{fs}}}{r_0^{\,2}\tau}-t \tag{5.31}$$

Therefore, it is possible to establish a graphical correlation with the left side of the equation using the linearization of time, which conduces to the Baker–Lonsdale equation:

$$f_t = \frac{3}{2}\left[1-\left(1-\frac{M_t}{M_\infty}\right)^{2/3}\right]-\frac{M_t}{M_\infty} = Kt \tag{5.32}$$

where K is the release constant, corresponding to the angular coefficient of the curve (inclination).

This equation can be utilized for linearization of release data for many microparticle formulations (microcapsules and microspheres).

5.9 Hopfenberg model

This is a model utilized to explain the drug release from erodible polymers and various geometrical forms. Hopfenberg proposed a series of equations to describe the drug release from planar, spherical or cylindrical films, with heterogeneous erosion (Katzhendler, Hofman, Goldberger, & Friedman, 1997):

$$\frac{M_t}{M_\infty} = 1-\left[1-\frac{k_0 t}{C_0 a_0}\right]^n \tag{5.33}$$

where M_t is the amount of the drug released at the time t, M_∞ is the amount released at the infinite time, $\frac{M_t}{M_\infty}$ is the fraction of the drug dissolved, k_0 is the erosion grade constant, C_0 is the initial concentration of the drug in the matrix, and a_0 is the initial radius of the sphere or cylinder or a half part of the thickness of the film. According to the geometrical form, the n value is 1, 2 or 3 for the film, cylinder or sphere, respectively.

With the aim to calculate the latency time (l) at the beginning of the drug release from the system, a modified equation was proposed by El-Arini and Leuenberger (1998):

$$\frac{M_t}{M_\infty} = 1-\left[1-k_l t(t-l)\right]^n \tag{5.34}$$

where k_1 is equivalent to k_0/C_0a_0.

Therefore, the limiting factors for drug release are matrix erosion and time, which depend on internal and external diffusing resistance.

Hopfenberg and Frish, based on the analysis of terms related to the stress of relaxation of the matrix and the diffusion flux, proposed a new approach:

$$M_t = M_{\infty,\mathrm{F}}\left[1-\frac{6}{\pi^2}\sum_{n=1}^{x}\frac{1}{n^2}\exp\left(-n^2 k_\mathrm{F} t\right)\right]+\sum_i^k M_{\infty,i}\left[1-\exp\left(-k_i t\right)\right] \qquad [5.35]$$

where $M_{\infty,\mathrm{F}}$ is the equilibrium on the absorbed amount during relaxation, k_{F} is the constant of the diffusion grade, $M_{\infty,i}$ is the amount that penetrates during the process of relaxation, and k_i is a constant related to the relaxation grade.

Therefore, this model shows separately the characteristics of diffusion and relaxation of permeability. Moreover, it is used to identify the mechanism of release from the optimized oilispheres using data derived from the composite profile, which essentially displayed site-specific biphasic release kinetics (Sibanda et al., 2004).

5.10 Weibull model

This model is a distribution function with the property to describe the phenomena and process associated to a finite time. It is based on a function proposed originally by Weilbull (1951), and Langenbucher (1972) adapted the function to describe the drug release curves.

The equation is expressed in terms of the drug fraction accumulated (m) in the solution on time t:

$$m = 1 - \exp\left[\frac{-(t - T_i)^b}{a}\right] \tag{5.36}$$

or

$$m = 1 - \exp^{-(t-T_i)^{\frac{b}{a}}} \tag{5.37}$$

The scale parameter (a) defines the timescale of the process. The localization parameter (T_i) represents the latency time of the release process, many times being zero. The form parameter (b) characterizes the type of curve:

(1) $b = 1$ (exponential)
(2) $b > 1$ (sigmoid, with ascendant curvature delimited by an inflection point)
(3) $b < 1$ (parabolic, displaying high initial slope and a consistent exponential character)

This model is considered more useful for comparing the drug release profiles of matrix systems. Considering the previous equation an empirical model, with no kinetic fundament, it is descriptive and does not allow the obtaining of inferences about the properties of the dissolution kinetics of the drug. There is no parameter of an intrinsic dissolution factor of a drug, and it does not allow the description of a type of *in vitro/in vivo* correlation.

5.11 Peppas–Sahlin model

This release kinetics model is a consequence of the indication that it is possible to calculate the approximate two contribution mechanisms (diffusional and relaxational) in an anomalous drug release process.

Considering the power law model and following an heuristic approach for the case of solvent transport in a polymer, first developed by Alfrey, Gurnee, and Lloyd (1966), diffusion and relaxation of polymeric chains can be considered additives.

Therefore, Peppas and Sahlin (1989) developed a release kinetics model that may write

$$\frac{M_t}{M_\infty} = K_1 t^m + K_2 t^{2m} \tag{5.38}$$

where K_1, K_2, and m are constants.

Considering the right side of the equation, the first term represents the Fickian diffusional contribution, F, whereas the second term represents the Case II relaxational contribution, R.

The coefficient m is the purely Fickian diffusion exponent for a system of any geometrical shape that exhibits controlled release (Peppas & Sahlin, 1989). The coefficient m is related to the coefficient n of power law, which was given for any shape, including films, cylinders and spheres.

When Equation (5.36) is compared to the power law equation (Equation (5.21)), m is equal to the coefficient n of power law. In this case, the relaxational mechanism is negligible, because the diffusional is expressed as a function of t^m and the relaxational contribution is expressed as a function of t^{2m}.

Figure 5.6 shows the variation of the diffusional coefficient m in relationship to the ratio of the diameter to the thickness of the system. The curve show two regions where the exponent is constant.

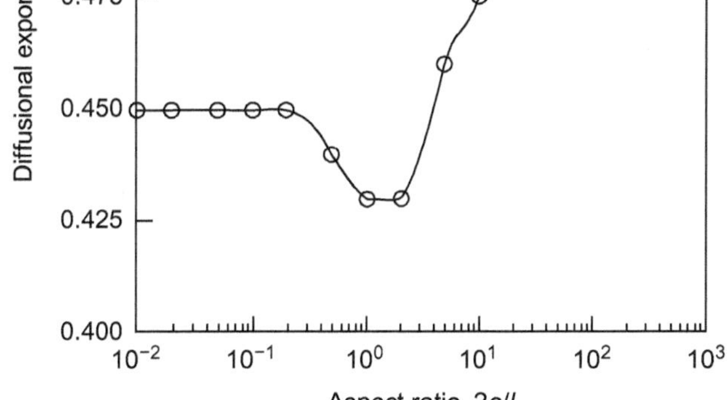

Figure 5.6 Variation of the Fickian diffusional exponent, m, with the aspect ratio, $2a/l$, where $2a$ is the diameter and l is the thickness (height) of the device.
Reprinted from Peppas and Sahlin (1989), with permission from Elsevier.

According to Figure 5.6, for the aspect ratio, *2a/l*, varying from 10^{-1} to 10^2, the value of coefficient m is 0.45 for $\frac{2a}{l} < 0.1$ and 0.50 for $\frac{2a}{l} > 100$.

Equation (5.38) can be written as

$$\frac{M_t}{M_\infty} = k_1 t^m \left[1 + \frac{k_2}{k_1} t^m\right] \tag{5.39}$$

Therefore, the amount of drug released due to the diffusion mechanism (Fickian, *F*) is calculated as

$$F = \frac{1}{1 + \frac{k_2}{k_1} t^m} \tag{5.40}$$

The ratio of both contributions can be calculated as follows:

$$\frac{R}{F} = \frac{K_2 t^m}{K_1} \tag{5.41}$$

When the analysis of the drug release mechanism uses a coupled diffusion/relaxation model, Peppas and Sahlin proposed the following methodology:

(1) Calculate the aspect ratio (diameter/thickness) of the system tested.
(2) Determine m from the values determined or from Figure 5.6.
(3) Plot the experimental release data and fit the first 60% to Equation (5.38).
(4) Calculate k_1 and k_2.
(5) Use Equation (5.38) to determine the percentage of Fickian and relaxational drug release.

To evaluate the drug release from a system, the fraction released (M_t/M_∞) on time t can be investigated comparing different models (e.g., Peppas–Sahlin, Korsmeyer–Ritger–Peppas and Higuchi mathematical models).

The release parameters, coefficient of determination (r^2) and Akaike information criterion (AIC) can be used to statistically analyze the results. The first 60% of all release curves are used for statistical analysis and, based on the statistically higher r^2 and lower AIC, it is possible to get the mechanism by which the drug release is governed. Therefore, the results indicate that the release of the active agent from the system is governed by the combination of Fickian diffusion and Case II relaxation.

For example, Zhang, Fub, Zhang, Xua, and Gao (2014) studied the model of the drug trimetazidine hydrochloride from systems composed of chitosan-oxidized glycerol monooleate in simulated gastric fluid (pH 1.2). The formulations in which the active agent may undergo diffusion out of the polymer, having the properties of swelling and chain relaxation, with the fraction released (M_t/M_∞) on time t, were investigated according to the Peppas–Sahlin, Korsmeyer and Peppas (power law)

Table 5.2 **Results obtained by release investigation of trimetazidine hydrochloride from chitosan-oxidized glycerol monooleate hydrogels (IH2, IH7 and IH8) and statistical analysis**

Samples and release media	Ritger–Peppas model					Peppas–Sahlin model				Higuchi model		
	k	n	r^2	AIC	k_1	k_2	m	r^2	AIC	k	r^2	AIC
IH2 in pH1.2	9.30	0.41	0.992	21.56	7.39	2.28	0.30	0.994	20.05	6.77	0.9759	66.80
IH7 in pH1.2	10.38	0.40	0.983	28.90	8.37	2.37	0.29	0.987	28.55	7.28	0.967	72.79
IH8 in pH1.2	8.20	0.44	0.995	17.50	6.42	2.17	0.31	0.996	17.04	6.44	0.981	61.64

Reprinted from Zhang et al. (2014), with permission from Elsevier and Higuchi mathematical models. The first 60% of all release curves were used for statistical analysis. The release parameters, coefficient of determination (r^2) and AIC (Zhang et al., 2010) for determining the release mechanism were statistically analyzed (Table 5.2).

The authors observed that the release profiles of the hydrogels were better described by the Peppas–Sahlin mechanism, the combination of Fickian diffusion and Case II relaxation, based on the statistically higher r^2 and lower AIC.

References

Agnes, E. J., & Gonzales Ortega, G. (2003). Mathematical models and physicochemical of diffusion. *Pharmacy Book*, 19(1), 9–19.

Alfrey, T., Gurnee, E. F., & Lloyd, W. G. (1966). Diffusion in glassy polymers. *Journal of Polymer Science*, C12, 249–261.

Baker, R. W., & Lonsdale, H. S. (1974). Controlled release: mechanisms and rates. In A. C. Taquary, & R. E. Lacey (Eds.), *Controlled release of biologically active agents* (pp. 15–71). New York: Plenum.

Costa, P. (2001). An alternative method to the evaluation of similarity factor in dissolution testing. *International Journal of Pharmaceutics*, 220, 77–83.

Dash, S., Murthy, P. N., Nath, L., & Chowdhury, P. (2010). Kinetic modeling on drug release from controlled drug delivery systems. *Acta Poloniae Pharmaceutica*, 67(3), 217–223.

El-Arini, S. K., & Leuenberger, H. (1998). Dissolution properties of praziquantel-PVP systems. *Pharmaceutica Acta Helvetiae*, 73, 89–94.

Ford, J. L., Mitchell, K., Rowe, P., Armstrong, D. J., Elliot, P. N. C., Rostron, C., et al. (1991). Mathematical modeling of drug release from hydroxypropylmethylcellulose: Effect of temperature. *International Journal of Pharmaceutics*, 71, 95–104.

Higuchi, T. (1961). Rate of release of medicaments from ointments bases containing drugs in suspension. *Journal of Pharmaceutical Sciences*, 50(10), 874–875.

Higuchi, W. I. (1962). Analysis of data on the medicament release from ointments. *Journal of Pharmaceutical Sciences*, 51(8), 802–804.

Higuchi, T. (1963). Mechanism of sustained-action medication. *Journal of Pharmaceutical Sciences*, 52(12), 1145–1149.

Hixson, A. W., & Crowell, J. H. (1931). Dependence of reaction velocity upon surface and agitation: I–Theoretical consideration. *Industrial and Engineering Chemistry*, 23, 923–931.

Katzhendler, I., Hofman, A., Goldberger, A., & Friedman, M. (1997). Modeling of drug release from erodible tablets. *Journal of Pharmaceutical Sciences*, 86, 110–115.

Kim, H., & Fassihi, R. (1997). Application of binary polymer in drug release rate modulation. 2. Influence of formulation variables and hydrodynamic conditions on release kinetics. *Journal of Pharmaceutical Sciences*, 86(3), 323–328.

Klech, C. M., & Simonelli, A. P. (1989). Examination of the moving boundaries associated with non-Fickian water swelling of glassy gelatin beads: Effect of solution pH. *Journal of Membrane Science*, 43, 87–101.

Korsmeyer, R. W., Gurny, R., Doelker, E. M., Buri, P., & Peppas, N. A. (1983). Mechanism of solute release from porous hydrophilic polymers. *International Journal of Pharmaceutics*, 15, 25–35.

Kosmidis, K., Argyrakis, P., & Macheras, p. (2003). Fractal kinetics in drug release from finite fractal matrices. *Journal of Chemical Physics*, 119, 6373–6377.

Langenbucher, F. (1972). Linearization of dissolution rate curves by the Weibull distribution. *Journal of Pharmacy and Pharmacology*, 24, 979–981.

Langer, R. S., & Peppas, N. A. (1983). Chemical and physical structure of polymers as carriers for controlled release of bioactive agents: A review. *Journal of Macromolecular Science – Reviews in Macromolecular Chemistry and Physics*, C23(1), 61–126.

Mauger, J. W., Chilko, D., & Howard, S. (1986). On the analysis of the dissolution data. *Drug Development and Industrial Pharmacy*, 12, 969–992.

Narasimhan, B., & Peppas, N. A. (1997). The role of modeling studies in the development of future controlled-release devices. In K. Park (Ed.), *Controlled drug delivery* (pp. 529–558). Washington: ACS.

Noyes, A. A., & Whitney, W. R. (1897). The rate of solution of solid substances in their own solutions. *Journal of the American Chemical Society*, 19, 930–934.

O'Hara, T., Dunne, A., Butler, J., & Devane, J. (1998). A review of methods used to compare dissolution profile data. *Pharmaceutical Science and Technology Today*, 1, 214–223.

Omidian, H., & Park, K. (2010). Introduction to hydrogels. In R. M. Ottenbrite, K. Park, & T. Okano (Eds.), *Biomedical applications of hydrogels handbook* (pp. 1–16). New York: Springer.

Peppas, N. A. (1985). Analysis of Fickian and non-Fickian drug release from polymers. *Pharmaceutica Acta Helvetiae*, 60, 110–111.

Peppas, N. A., & Narasimhan, B. (2014). Mathematical models in drug delivery: How modeling has shaped the way we design new drug delivery systems. *Journal of Controlled Release*, 190, 75–81.

Peppas, N. A., & Sahlin, J. J. (1989). A simple equation for the description of solute release. III. Coupling of diffusion and relaxation. *International Journal of Pharmaceutics*, 57, 169–172.

Pillay, V., & Fassihi, R. (1999). In vitro release modulation from crosslinked pellets for site-specifric drug delivery to the gastrointestinal tract. *Journal of Controlled Release*, 59, 229–242.

Polli, J. E., Rekhi, G. S., Augsburger, L. L., & Shah, V. P. (1997). Methods to compare dissolution profiles and a rationale for wide dissolution specifications for metoprolol tartrate tablets. *Journal of Pharmaceutical Sciences*, 86, 690–700.

Ritger, P. L., & Peppas, N. A. (1987). A simple equation for describing of solute release. I. Fickian and non-Fickian release from non-swellable devices in the form of slabs, spheres, cylinders or discs. *Journal of Controlled Release*, 5, 23–36.

Seki, T., Kawaguchi, T., Endoh, H., Ishikawa, K., Juni, K., & Nakano, M. (1980). Controlled release of 3,5-diester prodrugs of 5-fluoro-2-deoxyuridine from poly-L-lactic acid microspheres. *Journal of Pharmaceutical Sciences*, 19, 985–987.

Sibanda, W., Pillay, V., Danckwerts, M. P., Viljoen, A. M., van Vuuren, S., & Khan, R. A. (2004). Experimental design for the formulation and optimization of novel cross-linked oilispheres developed for in vitro site-specific release of mentha piperita oil. *AAPS PharmSciTech*, 5, 18.

Siepmann, J., & Peppas, N. A. (2012). Modeling of drug release from delivery systems based on hydroxypropyl methylcellulose (HPMC). *Advanced Drug Delivery Reviews*, 64, 163–174.

Walters, K. A., & Brain, K. R. (2009). Dermatological formulation and transdermal systems. In K. A. Waletrs (Ed.), *Dermatological and transdermal formulations* (pp. 319–400). New York: Informa.

Weilbull, W. (1951). A statistical distribution function of wide applicability. *Journal of Applied Mechanics*, 18, 293–297.

Zhang, J., Fub, M., Zhang, M., Xua, L., & Gao, Y. (2014). Synthesis of oxidized glycerol monooleate-chitosan polymer and its hydrogel formation for sustained release of trimetazidine hydrochloride. *International Journal of Pharmaceutics*, 465, 32–41.

Zhang, Y., Huo, M., Zhou, J., Zou, A., Li, W., Yao, C., et al. (2010). DDSolver: an add-in program for modeling and comparison of drug dissolution profiles. *The AAPS Journal*, 12, 263–271.

Drug delivery systems

6

6.1 Matrix tablets

Over the years, drug delivery systems have become extremely important to medicines. The interest in matrix tablets is based on their pharmacokinetic and biopharmaceutical advantages in relationship to conventional dosage forms.

Therefore, a modified-release formulation may improve management of a disease, as a result of the therapeutic benefits of therapeutic agent enhancement and minimizing of side effects. In addition, the convenience of more adjustable or effective doses boosts patient compliance to treatment.

Peroral administration of active agents is the most preferred route, and tablets are the most common and widely used dosage form. It is well known that they provide an immediate release of the therapeutic agent, normally resulting in relatively rapid drug absorption.

The strategy to control the drug release using matrix tablets has provided a breakthrough for drug delivery systems and started with sustained release formulations. Nowadays, matrix tablets are defined as oral solid dosage forms containing the therapeutic agent homogeneously dispersed throughout the hydrophilic or hydrophobic matrices that control the release rate, mainly by the type and proportion of polymer used in the preparations.

The release of the active agent happens in a continuous and controlled manner. Physicochemical mechanisms, like dissolution and diffusion, are the main way to control the release of the actives. Considering the different solubility properties, the drug can be dispersed in swellable hydrophilic substances, insoluble, inert, or lipid matrices.

Therefore, the preparation of matrix tablets constitutes a feasible strategy to obtain a controlled-release system. It is composed by direct compression of the active agent and excipients (retardant material and additives) to form a tablet in which the active is dispersed in the matrix that controls the release. Alternatively, the active agent and the material of the matrix can be granulated before compression.

As discussed in the previous chapters, drug delivery systems can be classified into two main categories (reservoir and monolithic matrix), and the polymers are between the main materials that can be used. In addition, the tablets can receive the coating, constituting a reservoir system where the release of the active agent is controlled by the coating membrane as well. Due to these possibilities, matrix tablets offer some advantages compared with traditional tablet dosage forms (Banker & Anderson, 2001; Siepmann & Peppas, 2001; Dash & Verma, 2013):

- They can maintain the therapeutic concentration of the active agent over a prolonged time period, avoiding high-blood concentration.
- There is reduction in toxicity by slowing absorption of the therapeutic agent.

Strategies to Modify the Drug Release from Pharmaceutical Systems. http://dx.doi.org/10.1016/B978-0-08-100092-2.00006-0

- They can minimize local and systemic side effects.
- There is improvement in treatment efficacy and better utilization of the active agent.
- They can minimize accumulation of the therapeutic agent with chronic dosing.
- They can be made to release high molecular weight compounds.
- They increase stability by protecting the active agent from hydrolysis or other derivative changes in the gastrointestinal tract.
- There is a reduction in health care costs and in usage of the total active agent.
- There is improvement of patient compliance.

In contrast, matrix tablets display some disadvantages (Banker & Anderson, 2001; Siepmann & Peppas, 2001; Dash & Verma, 2013):

- Sometimes, the remaining matrix needs to be removed after the active agent has been released.
- The system shows greater dependence on residence time in the gastrointestinal tract.
- There is increased potential for first-pass metabolism.
- There is a delay in the onset of drug action, and the burst effect is necessary.
- Food and the rate of transit through the gut may affect the release rate of the active agent.
- A continuous decrease of the release rate due to increased diffusional resistance and a decrease in effective area at the diffusion front are observed.

Considering that tablets are constituted by matrices, these structures can be divided into hydrophilic (soluble) and hydrophobic (insoluble) matrices. Hydrophilic matrices are divided into those displaying unlimited swelling and those displaying limited swelling. The difference between them is in the type of swelling, which is related to kinetics and the drug release mechanism. Moreover, the type of swelling of these matrices is associated to the different characteristics of polymers used.

In general, matrix systems have been submitted to many classifications, according to matrix structure, release kinetics (e.g., zero and first order) and mechanism of release control (e.g., diffusion, erosion, and swelling). In addition, the chemical nature and properties of materials can also be used as criteria (Table 6.1).

The active agent will be released from the matrix considering this classification. Mineral matrices control the drug release by intermolecular forces (attractive) and diffusion. Hydrophilic matrices with unlimited swelling display the control of the drug

Table 6.1 Classification and characteristics of matrix systems used in tablets

Matrix	Characteristics
Mineral	Active agent retained or adsorbed in the support (drug release controlled by diffusion)
Hydrophilic	Unlimited swelling (drug release controlled by diffusion) Limited swelling (drug release controlled by limited swelling)
Inert	Drug release by diffusion
Lipid	Drug release by diffusion and/or by erosion
Nonlipid biodegradable	Implants and microparticulate systems (drug release controlled by environmental characteristics (e.g., pH and enzyme presence))

release by diffusion and polymer chain relaxation. The swelling will control the drug release from hydrophilic matrices with limited swelling. Inert matrices are characterized by the drug release being controlled by diffusion. Lipid matrices can control the drug release by diffusion and/or erosion. Finally, nonlipid biodegradable matrices can control the release of the active agent by their degradation (e.g., enzymatic and pH).

Among these types of matrices, the hydrophilic and inert ones are normally more common in tablet form. Thus, modified-release tablets are a group of matrix systems with the properties to prolong and/or control the release of an active agent dispersed in a platform resistant to desegregation (Buri, 1987). Nonlipid biodegradable matrices are destined for a parenteral route (e.g., implants, microparticles, and injections).

In addition, it is very important to distinguish between the homogeneous and heterogeneous matrices. Homogeneous matrices are constituted by a nonporous phase, in which the active agent diffuses into the support due to its great solubility, while in heterogeneous matrices, the dissolved active agent needs to be around the polymeric particles and diffuse by small spaces (pores or channels) (Figure 6.1).

Inert matrices are normally heterogeneous because their form is maintained due to the porous structure formed by the polymer. Hydrophilic matrices are normally homogeneous because they are prepared with some polymers and, after hydration, display neither channels nor impermeable particles. However, Patil, Dordick, and Rethwisch (1996) showed the morphology of a hydrophilic heterogeneous matrix of cross-linked poly(sucrose acrylate) immediately after preparation, without drying. The morphology of the matrix consisted of high-density microgel domains in a continuous low-density matrix. The channel sizes were dependent on the preparation technique. A more concentrated matrix resulted in smaller channels (Figure 6.2).

As a consequence of their components, the matrices used in tablets can be also organized into three categories (Table 6.2). The first category is composed of materials

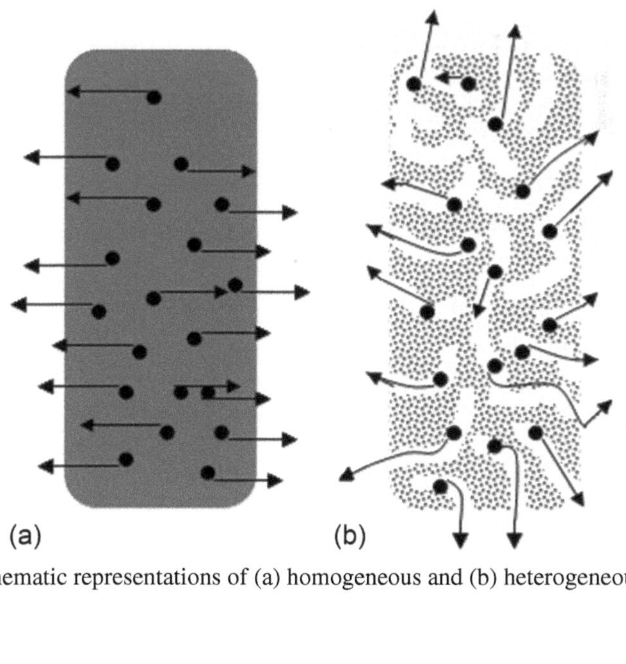

Figure 6.1 Schematic representations of (a) homogeneous and (b) heterogeneous matrices.

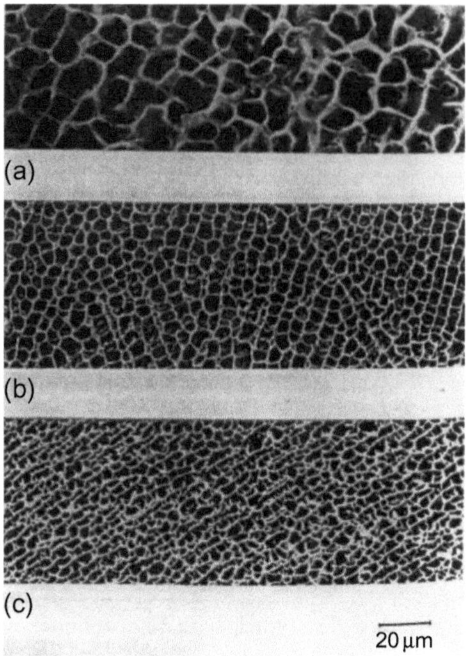

Figure 6.2 Scanning electron micrographs of poly(sucrose acrylate) hydrophilic matrices prepared with different initial monomer concentrations: (a) 15% (w/v), (b) 30% (w/v), and (c) 50% (w/v).
Reprinted from Patil et al. (1996), with permission from Elsevier.

Table 6.2 Categories of tablet matrix and examples of materials used to produce the tablets

Matrix	Characteristics in water	Examples of materials
Insoluble	Inert	Ethyl cellulose, methyl acrylate metacrylate copolymer, polyvinyl chloride and polyethylene
Insoluble	Erodible	Polyethylene glycol, polyethylene glycol monostearate, triglycerides, stearic acid, stearic alcohol, carnauba wax, and hydrogenated castor oil
Hydrophilic	Soluble	Cellulose derivatives (methyl cellulose, hydroxyethyl cellulose, hydroxypropylmethyl cellulose and sodium carboxymethyl cellulose) Non-cellulose polysaccharides (galactomannose, guar gum, alfarroba gum, sodium alginate, potassium alginate, agar, carrageen, xanthan gum, arabic gum, and caraia gum) Acrylic acid polymers (carboxypolymethylene)

that produce insoluble matrices, while the second category is composed by materials that are insoluble in water and potentially erodible. The third category is constituted by polymers that form hydrophilic matrices.

These materials are usually supplied in micronized forms, because small particle size is very important to an efficient dispersion and fast formation of a viscous layer on the tablet surface.

Moreover, the active agents that have to be added to matrix tablets should be orally effective and stable in the gastrointestinal tract medium. Drugs with short half-lives (2–4 h) are good candidates for formulation, as well as those for which the dose is up to 1.0 g and the therapeutic range is high and wide enough such that variations in the release do not result in concentration beyond the minimum toxic levels. Matrix tablets should not be prepared containing therapeutic agents that need large doses, present long biological half-lives, are very potent drugs and do not show a relationship between blood levels and biological activity (Banker & Anderson, 2001).

6.1.1 Hydrophilic matrices

Hydrophilic matrices can be classified into two groups:

- Classic hydrophilic matrices (unlimited swelling)
- Limited swelling matrices

There is a great difference between the two because of the controller mechanism of the drug release. Unlimited swelling matrices display diffusion as the main mechanism of drug release control. On the other hand, considering hydrophilic matrices with limited swelling, the mechanism of drug release control is the velocity of the dissolution medium (penetrant).

6.1.1.1 Unlimited swelling hydrophilic matrices

Unlimited swelling hydrophilic matrices are constituted by a mixture of one or more active agents and hydrophilic polymers. When in contact with water or biological fluids, eventual dissolution of a small fraction of the active agent can occur, followed by hydration and progressive gelling of macromolecules, forming a high-viscosity layer that increases in thickness with time. This gelled layer or barrier controls the penetration of water in the tablet center direction and also constitutes a barrier to fast release by diffusion of the active agent to the exterior.

The process and the velocity of release are dependent on physicochemical characteristics of the active agent, of the tablet and its hydration. In addition, the hydrophilic matrices with unlimited swelling display more advantages (Veiga, 1988):

- They are a simple technology to prepare.
- They use safe materials.
- They have low consistency when hydrated into the gastrointestinal tract.
- It is possible to add a great amount of active agent.
- The drug release is relatively independent of physiological variables.

There are many gelling substances used in the preparation of hydrophilic matrices, and their use is based on important characteristics such as nominal viscosity, behavior under compression, gelling velocity and water penetration, and compatibility with the other components of the formulation, particularly with ions.

Many formulation factors are susceptible to modifying the release kinetics of the active agent in these types of hydrophilic matrices:

- Concentration and viscosity of jellying agent.
- Compression force.
- Nature of diluent.
- Matrix acidification.

Cellulose derivatives are the substances more utilized in the preparation of this type of matrix. Particularly, hydroxypropylmethyl cellulose (HPMC) is compatible with electrolytes, and its viscosity is not changed by pH. In relationship to compression capacity, HPMC and hydroxyethyl cellulose (HEC) can be directly compressed. The matrices formed by methyl cellulose (MC) are not very resistant, because the gelled layer does not adhere to the tablet. The gums display many variations of inter batches, while the gelling of carbomers occurs by neutralization of the acid form, indicating that the pH variations of the gastrointestinal tract affect their viscosity.

Hydrophilic matrix tablets of unlimited swelling are obtained like the conventional tablets and use the traditional adjuvants. When it is necessary to use granulation, it needs to be performed using alcoholic solutions, avoiding the use of water, which would cause undesirable gelling.

Considering the hygroscopic characteristics of matrix forming substances, the tablets need to be prepared under low-relative humidity and packed into impermeable packaging.

6.1.1.2 Limited swelling hydrophilic matrices

These types of matrices have shown great development, considering they present release kinetics of zero order, or very near. Moreover, they display some similarities with hydrophilic matrices with unlimited swelling, because the velocity of penetration of the medium (penetrant) in the matrix is the limiting step. They are called "activated by solvent."

Another characteristic is in relationship to the polymer, which, in the anhydrous state, needs to be vitreous at body temperature and relax its molecular structure when in the presence of water, to allow a quantitative diffusion of the active agent through the gelling layer.

The swelling and solubility of the polymer in water are the other important characteristics. The materials display very slow hydration and limited swelling (100% of initial weight), but they are insoluble in an aqueous environment. These characteristics are possible due to the presence of physical and/or chemical bonds in the polymer structure.

Therefore, hydrophilic matrices with limited swelling possess advantages and disadvantages. The main advantages are

- Release kinetics are approximately of zero order.
- They are easy to prepare (using a conventional preparation process).

- They enable the addition of active agents of easy degradation.
- Some matrices display bioadhesive properties.

On the other hand, the drawbacks are

- Zero-order kinetics are not possible with a high active agent amount.
- The active agent can react with the monomers or with cross-linking agents.
- The active agent may be degraded when it is necessary to use high temperatures during preparation (cross-linking or drying process).

There are many polymers utilized the preparation of limited swelling hydrophilic matrices (Table 6.3), considering that they need to be in the vitreous state at body temperature, and they must be insoluble, undergo limited swelling in an aqueous medium, and be nontoxic.

The preparation of this type of tablet has two steps. First, the active agent is added in the polymer; then, the form to the matrix is accomplished. The incorporation of the active agent may be performed in three ways:

(1) The active agent is dispersed into the pre-prepared polymer dispersion and the incorporation is obtained by polymerization and posterior cross-linking.
(2) The polymer is impregnated by a dispersion of the active agent. The advantage of this method is to allow a previous purification of the polymer. In addition, the incorporation may be limited by drug solubility in solvents able to swell the polymer before the free spaces in the macromolecular net can be reached.
(3) Thermoplastic polymers enable the incorporation of the active agent by the mixture with the powder polymer. The incorporation of the active agent in the macromolecular net occurs during the hot preparation, and this technique is restricting to thermostable drugs.

Generally, the tablets are prepared by compression of particles obtained by granulation of bulk prepared with the polymer–drug mixture.

6.1.2 Lipid matrices

Lipid matrices can be manufactured using hot melt granulation, roller compaction or standard direct compression. The preparation involves the mixture of powered components (wax matrix former, channeling agent, solubilizer, pH modifier, anti-adherent/glidant, and lubricant) and the addition of an active agent, which will be contained in the hydrophobic matrix that remains intact during drug release.

The use of hydrophobic materials in pharmaceutical sciences has received great attention due to the clinical and economic benefits that they can achieve (Abu-Diak, Andrews, & Jones, 2012). The hydrophobic matrix formers form 20–40% of the formulation and need to be solid at room temperature and not melt at body temperature. The most used materials are hydrogenated vegetable oils, soy oil, cottonseed oil, microcrystalline wax, and carnauba wax.

Channeling agents are substances that are soluble in the gastrointestinal tract and leach from the formulation, leaving tortuous capillaries through which the dissolved active agent may diffuse in order to be released. It is important to note that the active agent can exert the function of the channeling agent, but it needs to show suitable

Table 6.3 **Examples of limited swelling polymers**

Polymer	Monomer
Polyvinyl alcohol (PVA)	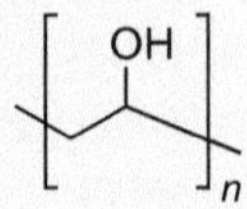
Poly(hydroxyethyl methacrylate) (PHEMA)	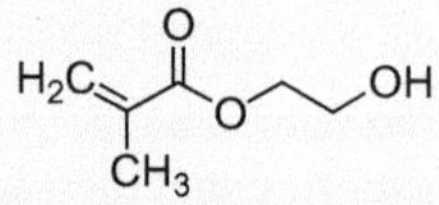
Ethylene vinyl acetate (EVA)	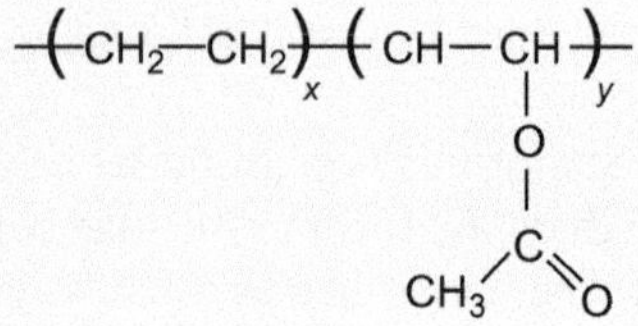
Polyethylene oxide (PEO)	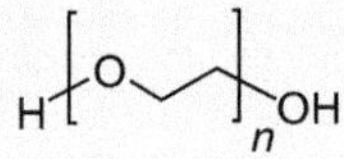
Polyacrylonitrile (PAN)	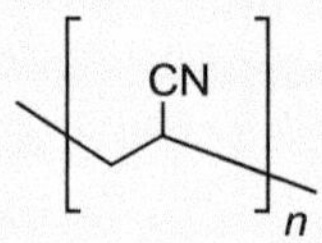
Polyacrylamide (PAA)	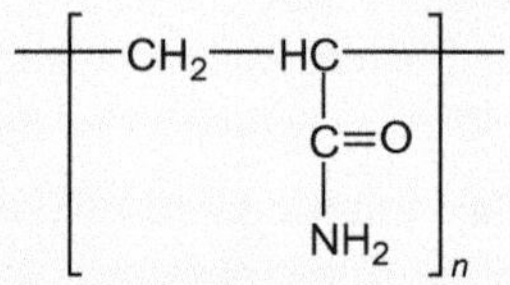
Polyvinyl acetate (PVAc)	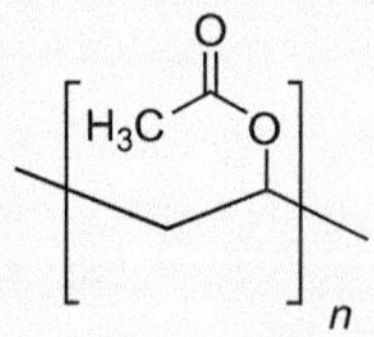
Poly(hydroxyethyl methacrylate-co-methyl methacrylate) (P(HEMA-co-MMA))	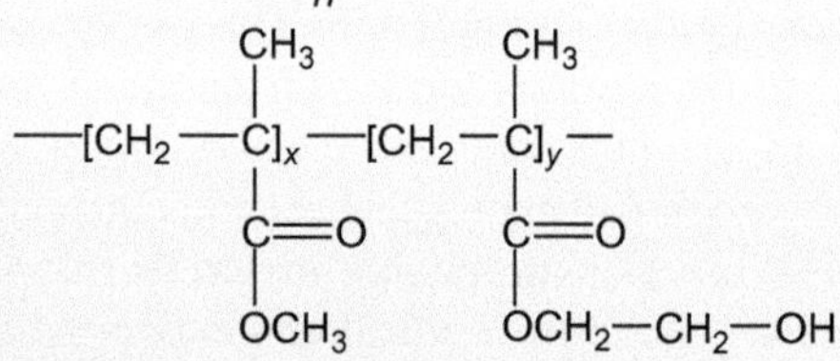
Poly(hydroxyethyl methacrylate-co-N-vinylpyrrolidone) (P(HEMA-co-NVP))	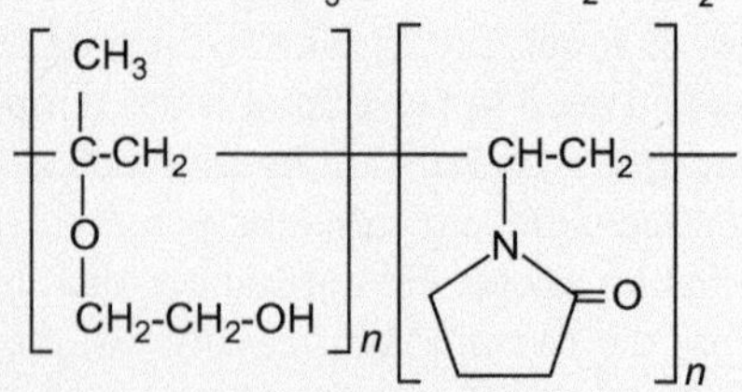

solubility in water. The choice depends on the drug, the desired release characteristics and the amount of the material that needs to be present (about 20–30%). Examples of channeling agents are sugar, sodium chloride, and polyols.

Solubilizers and pH modifiers are often used to enhance the dissolution of the active agent. The aim can be achieved using polyethyleneglycols, polyols, or surfactants. When the active agent is ionizable, the use of counter-ions or buffers is appropriated. These substances can also be the channeling agent.

During the compaction process, the hydrophobic matrix former may melt due to the heat generated. This phenomenon can cause sticking of the matrix to the punches. To resolve this problem, anti-adherents such as colloidal silicon dioxide (0.5–1%), magnesium stearate (0.25–5%), and/or talc (4–6%) must be used (Rowe, Sheskey, & Quinn, 2009). These materials have the ability to improve the flow of the preparation on the tablet machine.

The drug release is dependent on the channeling agent, which dissolves on an aqueous medium and leaches out of the compact, forming a porous matrix of tortuous capillaries (Collett & Moreton, 2001). Therefore, a way of the water-filled capillaries is formed and, when the active agent dissolves in the aqueous medium (penetrant), it diffuses out of the matrix.

Considering porosity, matrix tablets may also be classified in macro, micro, and nonporous systems (Munday & Cox, 2000). Macroporous systems display drug diffusion through the pores or channels of the matrix, which range from 0.1 to 1 μm in size (pore size is larger than diffusant molecule size). In microporous systems, diffusion occurs essentially through pores, which are of size range 50–200 A (slightly larger than diffusant molecule size). Finally, in nonporous systems, the molecules of the active agent diffuse through the network meshes (only the polymeric phase exists and no pore phase is present).

Lipid matrices are a simple kind of delivery system displaying control of rate and extension of drug release. Considering the previous description, the drug release is generally not zero order, and the Higuchi equation (square root of time release kinetics) can be used to describe the corresponding release characteristics.

Moreover, hydrophobic matrices generally are not suitable for insoluble active agents because the concentration gradient is too low to render adequate drug release (Munday & Cox, 2000).

6.1.3 Mechanism of release of therapeutic agent

When the matrix tablet reaches the aqueous medium of the body (e.g., gastrointestinal fluid), the active agent in the outside layer is dissolved first and then diffuses out of the matrix. It is a continuous process that involves the interface between the medium solution and the solid active agent moving toward the interior.

For lipid matrices, hydrophilic unlimited swelling matrices and other diffusion-controlled systems, the rate of dissolution of active agent particles within the matrix must be faster than the diffusion rate of the dissolved active leaving the matrix. Considering that a pseudo-steady state is maintained during drug release, the diameter of the drug particles is less than the average distance of drug diffusion through the matrix, and the bathing solution provides sink conditions at all times, the release behavior for the matrix tablet system can be mathematically described by a plot of the drug release versus square root of time, which will result in a straight line, as discussed in Chapter 5.

In a monolithic matrix containing pores or channels, the drug release process is due to the simultaneous penetration of liquid, dissolution and leaching out of the active agent through tortuous interstitial spaces. The length and volume are very important, and the equation of the drug release can be described as

$$\frac{Q}{t^{0.5}} = S\left[D_s \cdot C_s \cdot p \cdot \left(2Q_0 / V\right) - p \cdot C_s\right]^{0.5} \tag{6.1}$$

where $Q/t^{0.5}$ is the released amount; S is the effective surface of diffusion; D_s is the diffusion coefficient in the hydrated matrix; C_s is the solubility of the active agent in the release medium; p is the matrix porosity; T is the diffusional path length; C_0 is the initial concentration of the active agent; and t is the tortuosity of interstitial space.

According to Higuchi (1961, 1962, 1963), for the purpose of data treatment, Equation (6.1) can be reduced to

$$Q = k \cdot t^{0.5} \tag{6.2}$$

where Q is the amount released, k is the constant of release, and t is the time.

The anomalous release can be observed in some matrix systems. First, the release is controlled by diffusion, the matrix polymer may become hydrated and begin to swell, leading to release by relaxation of polymeric chains and erosion (unlimited swelling). It is a complex mechanism in which the mathematical model is based on the changes in diffusional path length. The penetration of the liquid medium into the matrix tablet conduces to the formation of a series of fronts that later disappear along the process of matrix dissolution (Figure 6.3) (Veiga, 1988; Banker & Anderson, 2001; Ferrero, Massuelle, Jeannerat, & Doelker, 2008, 2013; Ferrero et al., 2010).

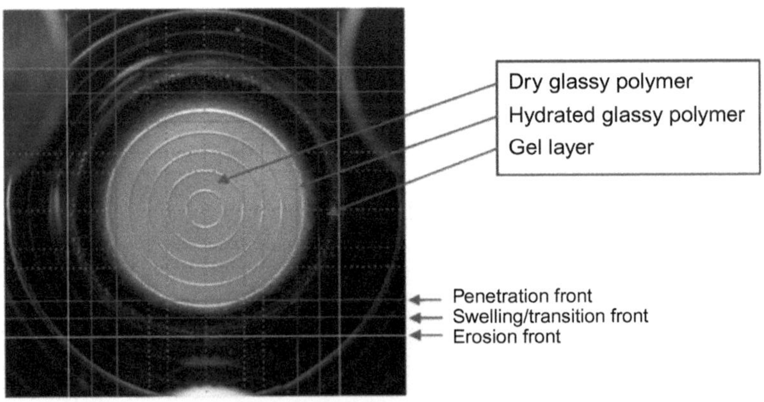

Figure 6.3 Photograph of the tablet (fitted into the polymethacrylate device used for radial drug release and swelling experiments) after a 6-h immersion in water added with methylene blue. The photograph shows the penetration front (pink lines), the swelling/transition front (green lines) and the erosion front (yellow lines). The dotted blue lines are reference lines from which the original photograph was adjusted.
Reprinted from Ferrero, Massuelle, and Doelker (2010), with permission from Elsevier.

Ferrero et al. (2010) described forefront movements: the penetration front (dry glassy/glassy hydrated polymer interface), showing that the solvent concentration is near zero beyond the glassy/rubbery interface. There is the entry of the aqueous medium into the matrix, swelling of the matrix, dissolution of the active agent in the liquid, diffusion of the active through the gel layer, and erosion of the swelled matrix (Ferrero et al., 2010). Therefore, the concentration, viscosity, and chemical structure of the rubbery polymer are important parameters in the matrix performance, because they determine the strength of the gel layer. Some hydrophilic polymers are not suitable for preparation of swellable matrices because they do not form the gel layer quickly. Carboxymethyl cellulose, hydroxypropyl cellulose, and tragacanth gums are examples (Rowe et al., 2009; Dash & Verma, 2013).

Non-biodegradable matrices control the release by diffusion of the drug through the gel layer, resulting in first-order release kinetics.

6.2 Suspensions

When two substances displaying different physical states are put in contact, there is the formation of two phases, and a determined surface of contact between them is formed. As one substance decreases in size, the surface area increases, there is the distribution of this phase into the other, and the preparation tends to be increasingly homogeneous.

Depending on the size of the particles of substances, they will be formed from coarse to thin dispersions (molecular or ionic dispersions). Solutions are generally defined as homogeneous mixtures of two or more substances, at molecular or ionic level (size less than 0.001 μm), forming a single phase. Colloidal dispersions are characterized to possess dispersed particles with size from 0.001 to 0.5 μm, spanning the gap between true solutions and coarse dispersions. In a coarse dispersion, one can usually see the particles of the dispersed phase either with the eye or with the optical microscope. The transition from colloidal to coarse dispersions is gradual, and the theoretical concepts that govern colloidal dispersions can also be applied to the larger particles found in coarse dispersions (Falkiewicz, 1996).

Depending on the state of each phase, the coarse dispersion can be formed by liquid or solid dispersed in gas; gas, liquid, or solid dispersed in solid; or gas, liquid, or solid dispersed in liquid.

Pharmaceutical suspension may be defined as a coarse dispersion in which finely divided solid drug particles (the *suspensoid*) are dispersed in a second phase, generally an aqueous liquid more or less viscous, in which the drug is insoluble, rendering a final preparation of homogeneous appearance. An aqueous suspension is a useful formulation system for administering an insoluble or poorly soluble active agent (Atwood, 2001).

Liquid dispersion systems possess, among the various pharmaceutical dosage forms, the most complex physicochemical characteristics. This is due to the presence of two different phases in the formulation. The size of dispersed solid particles ranges from 0.1 to 100 μm, but they are generally greater than 1 μm in diameter. Solid particles can have various morphologies, and the most common shapes are spheres, cylinders, rods, needles, and various crystalline shapes. These physical characteristics of the solid phase are important to determine the physicochemical characteristics of the suspension.

Suspension dosage forms can be developed to be administered by oral, parenteral (e.g., intramuscular and subcutaneous), rectal, topical (conventional and in reservoirs in transdermal patch preparations) or ocular routes, and there are many factors that justify their usage (Allen, Popovich, & Ansel, 2011; Atwood, 2001; Jones, 2008):

(1) Easy administration (For many patients, such as the elderly, children, and infants, the liquid dosage form is preferred to the solid of the same active agent, because of the ease of swallowing liquids and the flexibility in administration of a range of doses.)
(2) The active agent is insoluble in the dispersing medium.
(3) Taste masking (The disagreeable taste of certain active agents in solution form is overcome when they are administered as undissolved particles of an oral suspension.)
(4) The protection or the increasing of the stability of the active agent (Some drugs are chemically unstable in solution, but stable when suspended.)
(5) They may be formulated to provide controlled drug delivery (e.g., intramuscular injections).

The rate of absorption of an active agent decreases in the following order: solution > suspension > capsules > tablets > coated tablets (Nash, 1988). The prolonged clinically useful plasma concentration of an active agent can be a result of the delayed absorption of this active.

However, the particles dispersed possess greater size, displaying a greater tendency to separate from the dispersion medium. Thus, the pharmaceutical suspensions are unstable, requiring formulation skill to ensure that the physical stability of the formulation is retained over the period of the shelf life. In addition, the formulation of aesthetic suspension formulations is difficult, and the formulations may be bulky and difficult for a patient to carry (Jones, 2008). Another disadvantage is the adhesion of suspension particles to container walls, particularly with low-dose drugs.

Thus, an acceptable suspension needs to possess certain desirable qualities:

- The suspended material should not settle too rapidly.
- Considering that it is essential to the accurate administration of uniform doses of the active agent, complete and uniform redistribution of the solid dispersed phase needs to be performed before use. This should be accomplished by moderate agitation of the container.
- The suspension must not be too viscous to pour freely from the bottle or to flow through a syringe needle.
- The deliberate flocculation needs to be avoided in the case of ophthalmic administration, where aggregation leads to agglomerated particles that do not easily pass through the drainage ducts of the eye.
- The formulation needs to be aesthetically pleasing.

Nowadays, suspensions of micro or nanoparticles and liposomes are developed using a variety of polymers, phospholipids, and/or proteins and constitute a new class of dispersed systems, which will be discussed further.

6.2.1 Sedimentation rate of the particles

Suspension is a thermodynamically unstable system, which should have a half-life long enough to facilitate its transport and storage in different conditions of temperature, humidity, and so on. Therefore, a pharmaceutical suspension should settle slowly and should be readily redispersed upon gentle shaking of the container. To control these characteristics, it is necessary that the size of the particles remains fairly constant throughout long periods of undisturbed standing, and that there is no crystal growing. Moreover, the rheological characteristics of formulation can be optimized to accomplish these necessities and to enable the suspension to pour readily and evenly from the container (Allen et al., 2011).

The nature of the dispersion medium, the dispersed phase, and pharmaceutical adjuncts are very important to determine these main features of a suspension. Moreover, gravitational force is also involved to produce the sedimentation of the dispersed phase. In this context, Stokes' equation (Equation 6.3) was derived for an ideal condition in which the perfectly spherical solid particles are in a very diluted suspension and they settle without turbulence, without colliding with other suspensoids, and without chemical or physical attraction or affinity for the dispersion medium (Allen et al., 2011).

$$\frac{\mathrm{d}x}{\mathrm{d}t} = \frac{d^2\left(\rho_\mathrm{i} - \rho_\mathrm{e}\right)g}{18\eta} \tag{6.3}$$

where $\mathrm{d}x/\mathrm{d}t$ is the settling rate; d is the diameter of the solid particles; ρ_i is the density of the solid particles; ρ_e is the density of the medium; g is the gravitational constant; and η is the viscosity of the medium.

The sedimentation rate is dependent on these factors, which can be adjusted to enhance the physical stability of the suspension. For example, as the particle size decreases, the settling rate decreases as well. The lower difference between the densities of both phases conduces to the reduced settling rate. Moreover, the increasing of the viscosity of the medium conduces to the decreasing of the settling rate. However, a suspension displaying too high viscosity is not generally desirable, because it is then difficult to pour and redisperse the suspensoid. Therefore, it is important to optimize the rheological characteristics of the suspension. This could be accomplished considering the medium used and the solid content. As the proportion of solid particles in a suspension increases, so does the viscosity.

The physical stability of pharmaceutical suspensions is most appropriately adjusted by an alteration in the dispersed phase rather than through great changes in the dispersion medium. Thus, the adjustments are focused on mainly uniformity of particle size, particle size, and separation of particles (Allen et al., 2011).

6.2.2 The dispersed phase

Particle size is very important to consider. According to Stokes' equation (6.3), it is beneficial to reduce the size of particles, because the sedimentation rate is reduced as the particles decrease in size. Actually, reduction in the size of suspensoids produces slow, more uniform rates of settling. On the other hand, it is necessary to avoid reducing the particle size too much, because fine particles have a tendency to form a compact cake upon settling to the bottom of the container, making the redispersion and homogenization of a settled suspension difficult.

As previously discussed, the best pharmaceutical suspensions have a particle diameter from 1 to 50 μm. Micropulverization, jet milling, and micronizing are used efficiently to reduce powders to the acceptable size. Particles of extremely small dimensions may also be produced by spray drying.

The shape, symmetry, and charge of particles can also affect the stability of the suspension. Asymmetric particles, especially those that are elongated, are more satisfactory than spherical ones (Atwood, 2001).

When the particles are charged, the surface charge density generated must be compensated by an opposite charge that will be distributed in the solution surrounding the solid particle. Therefore, the surface charge, together with its countercharge, constitutes the electrical double layer. The zeta potential is a measurable indication for the apparent particle charge in the dispersion medium.

A solid material dispersed in a liquid medium displays some important properties that are dependent on the existence of a surface charge (zeta potential). The origin of this surface charge is dependent on the nature of two phases (liquid and solid). The most frequent mechanisms are differential ion solubility, direct ionization of surface groups (a common mechanism in ionizable polymeric surfaces containing carboxylic or sulfonic acids, sulfuric acid esters, or amino and quaternary ammonium groups), isomorphous ion substitution (a common mechanism in many minerals, such as clays and oxides, in which ions of different valency substitute structural ions from the corresponding lattice ions), specific ion adsorption (e.g., adsorption of surfactants) and anisotropic crystals (some crystalline materials, like aluminosilicate layers in clays, have the important characteristic that when cleaved, they produce both positively and negatively charged surfaces).

When the zeta potential value is relatively high (positive or negative), the repulsive forces usually exceed the attractive forces. When the charged particles are together, the electrical repulsive forces between them allow them to slip past one another to form an organized arrangement at the bottom of the container. The big particles settle first and the small particles fill the voids, forming the "cake," sediment difficult to redisperse. The supernatant liquid may remain cloudy after sedimentation as the consequence of the presence of colloidal particles that remain dispersed. The weight of the particles conduces to a phenomenon in which those particles lowermost in the sediment are gradually pressed together, overcoming the repulsive barrier. Moreover, the physical bonding can lead to the formation of cake, and the formation of bridges between the particles resulting from crystal growth and hydration effects may also occur, with forces greater than agitation usually required to disperse the sediment.

A cake can be avoided by preventing agglomeration of particles into larger crystals or into masses. If the shape of the particles is favorable to accommodation and good organization of sediment, the particles will form a good organized structure and the interactions between their surfaces can be greater, increasing the difficulty of redispersing the particles. This rigid cohesion can be prevented by intentional formation of a less rigid or loose aggregation of the particles, held together by comparatively weak particle-to-particle bonds (Allen et al., 2011).

However, when the particles are flocculated, the sediment is formed fairly rapidly, it is loosely packed and with high volume, and the suspensoids are easily resuspended. In this sense, the supernatant liquid remains clear, considering that the colloidal particles are trapped within the structures and settle together.

Therefore, flocculated sediment is desirable for a pharmaceutical suspension. The flocculation rate is dependent on the number of particles present, so that the greater the number of suspensoids, the more collisions there will be, and the more flocculation is likely to occur. However, it is necessary to control the system with the aim to obtain a satisfactory flocculation state. The addition of electrolytes or ionic surface-active agents may lead to a controlled flocculation.

Flocculated particles can be obtained by several methods; the choice is dependent on the type of suspension desired. In addition, it is important to work a constant, or narrow, pH range because the magnitude of the charge on the drug particle can vary greatly with pH (Atwood, 2001).

The suspensoids may also be stabilized in the absence of a charge by use of nonionic polymeric materials (steric stabilization), which may increase the viscosity of the dispersion medium and thus slow the sedimentation rate. In addition, these materials will also form adsorbed layers around the particles so that the approach of their surfaces and aggregation to the coagulated state is avoided. For oral suspensions of active agent, clays (e.g., bentonite magma) are commonly used as a flocculating agent. In a parenteral suspension, where clays cannot be used, the flocculated particles can be obtained by an alteration in the pH of the formulation (generally to the region of minimum drug solubility).

It is common to use electrolytes as flocculating agents due to their properties of reducing the electrical barrier between particles and forming a bridge to link them together. Surfactants can also induce flocculation of particles in suspension, but their concentration needs to be carefully determined due to the possibility of destabilization of the system.

There is an equation to determine the sedimentation volume (F):

$$F = \frac{V_s}{V_i} \tag{6.4}$$

where V_s is the volume of the sediment and V_i is the initial volume of the suspension.

The magnitude of F may range from less than 1 to values that are greater than 1. In the case of deflocculated suspensions, the value of F is usually small, whereas the F for flocculated systems is high (i.e., close to or greater than 1) due to the large volume occupied by the flocculated structure. It is also possible to determine the degree of flocculation (the ratio of the ultimate sedimentation volume of the flocculated suspension to the ultimate sedimentation volume of the deflocculated suspension), which is usually the preferred measurement, as it provides a point of reference (e.g., the suspension before and after flocculation).

6.2.3 The dispersion medium

The particles that constitute the dispersed phase are dependent on the dispersion medium. The first point that needs to be considered is the possibility of the particles to increase their solubility in the medium, which is not good. Moreover, the dispersion medium needs to enable good rheological properties. For example, when the particles of suspension are flocculated, the velocity of settling is high and the suspension displays inelegant characteristics. The rapid formation of sediment hinders accurate measurement of dosage, and from an aesthetic point of view produces too unsightly a supernatant layer. To correct this problem, suspending agents (e.g., carboxymethylcellulose, methylcellulose, microcrystalline cellulose, polyvinylpyrrolidone, xanthan gum, and bentonite) are added to the dispersion medium to rheologically structure it (Allen et al., 2011). Basically, they thicken the dispersion medium and help suspend the particles.

These materials that constitute the suspending agents should not interfere (even partially) with the availability of the active agent. These agents can bind certain drugs, rendering them unavailable or only slowly available for therapeutic function. This phenomenon can be used to modulate the drug release.

In addition, the suspension cannot be too viscous with the addition of suspending agents, because the suspensoid redistribution can be difficult. Rheological studies may be performed in order to investigate the continuous flow properties of the dispersion medium, analyzing its consistency index, flow index, hysteresis area presence, and yield value.

The ability of the dispersion medium to support the active agent amount while maintaining desirable features of flow rheology must be considered when the solid content of a suspension is determined. Moreover, the dose of active agent and the volume to be administered must also be considered.

6.2.4 Preparation of suspensions

For the preparation of suspensions, the characteristics of both the dispersed phase and the dispersion medium need to be acquainted. Once the particles have been wetted, they must be separated and distributed uniformly throughout the liquid of the suspending medium. The degree of dispersion is the extent to which this is accomplished. A good degree of dispersion is displayed by particles that are broken down into primary particles without the existence of many twins or aggregates, indicating that particles are in a state dominated by individual particles, wetted and distributed in a uniform way.

Sufficient and efficient agitation of the mixture of two phases must be provided initially to obtain a high degree of dispersion, considering that wetting is favorable. Sometimes, the particles have an affinity for the suspending medium and are readily wetted by it. Other substances are not penetrated easily by the dispersion medium and have a tendency to agglutinate or float on the dispersion medium. In this case, the solid must first be wetted to make it more penetrable by the dispersion medium. This is possible with the use of wetting agents like alcohol, propylene glycol, glycerin, and other hygroscopic liquids. Their function is to displace the air in the crevices of the particles, dispersing them, and allowing penetration of the suspending medium into the solid phase.

The dispersion medium can receive all of the formulation's soluble components (e.g., colorants, flavorants, preservatives, sweeteners, water-compatible liquids, and opacifiers). It is important to prepare a dispersion medium that can support the particles to be suspended. Considering Stokes' law (Equation 6.3), the particles in suspension will exhibit a tendency to settle or rise. Therefore, it is necessary to use approaches to resist this natural tendency, such as reducing the particle size, lowering the density difference between the two phases ($\rho_i - \rho_e$) or optimizing the rheological characteristics of the suspending medium (e.g., increase the viscosity). Generally, some combination of these three approaches is employed. When the suspending medium is aqueous, it is possible to use hydrocolloids or some other type of suspending agent. They are preceded by the addition of the formulation's soluble components, particularly if they are difficult to dissolve (Falkiewicz, 1996). In addition, the suspending agent can be a finely divided insoluble material (e.g., clays or cellulose derivatives). After this, it is added in portions to the powder. Finally, a mixing device, a colloid mill or other blender is used to ensure uniformity.

The entrapment of air when adding the components of a suspension to the dispersion medium needs to be avoided. The presence of air entrapped may impart increased viscosity to a suspension initially and can lead to various potential problems (e.g., inaccurate dosage of active agent for medication that is dispensed by volume, problems of rheology, separation, and color inconsistencies).

Finally, the formulation should be evaluated for evidence of foam, undispersed particles, and settling.

Sometimes, the suspension needs to be an extemporaneous preparation in which the solid components of preparation are dried and water is added just before the use of formulation. This can be justified by the difficulty that confronts the pharmacist with the lack of ready information on the stability of an active agent in a liquid vehicle. It is known that many active agents in liquid form have faster decomposition rates than in solid form, and some are affected by the pH of the medium (Falkiewicz, 1996).

Extemporaneous preparation of suspensions of a drug available commercially only in other dose forms is widely practiced in hospital pharmacies, particularly for pediatric use. It is important to ensure that the suspending agent is readily and uniformly incorporated in the preparation, readily dissolved or dispersed in water without resorting to special techniques, and is inert, nontoxic, and free from incompatibilities. Moreover, it is necessary to certify the formation of a loosely packed system that does not cake, and that the preparation does not influence the dissolution rate or absorption rate of the active agent (Florence & Attwood, 2006).

6.2.5 Modified release of active agent

The large surface area of dispersed solid particles ensures a high availability for dissolution, and hence absorption (Atwood, 2001). However, the suspension dosage form displays properties that induce modified release of the active agent. First, considering the insolubility of the drug in the suspending medium, difficult of its release occurs.

When the dispersion medium is hydrophilic, the active agent has hydrophobic characteristics and its solubility in body fluids is low, resulting in low bioavailability. In addition, the hydrophobic dispersion medium makes the interactions of the hydrophilic drug and the body fluids difficult.

More reproducible drug release profiles can be obtained using suspensions containing a large number of particles. In addition, this may result in more predictable gastric emptying, more reproducible absorption, and minimized risk of dose dumping or local irritation (Yuan et al., 2014).

The formulation of suspensions for modified drug release has displayed sustained-release capabilities (Chang, 1992). Researchers have developed strategies like coated particles, drug-impregnated matrices, microencapsulation or ion-exchange resins. Extensive attention has been given to developing liquid oral sustained-release suspensions based on multiparticulates, such as coated pellets or microparticles (Yuan et al., 2014).

The main problem for developing these formulations is the difficulty of maintaining the stability of sustained-release particles in the dispersion system.

The combination of particle coating with ion-exchange resin complex resulted in ionic drugs complexed with ion-exchange resins, and the drug–resin complex particles were coated with ethylcellulose (Chang, 1992). The coated particles in suspension maintain the active agent adsorbed onto the resin, and its release is slowed by the ion-exchange process in the gastrointestinal tract. The antitussive and allergy relief formulation called Tussionex (Pennkinetic®) is an example of extended-release suspension in which hydrocodone polistirex and chlorpheniramine polistirex are in suspension and are released slowly, providing up to 12 h of relief per dose.

6.3 Osmotic pumps

Considering the advantages of drug delivery systems, a great number of advancements have been made in the development of new techniques and strategies. As previously discussed, the main aims are to regulate the rate of delivery of the active agent, sustaining the duration of therapeutic action and/or targeting its delivery to a specific tissue.

These advances conduced to the development of osmotic pumps, which are a form of a membrane-controlled release drug delivery system that uses osmotic pressure as the energy source. The fundamental aspect of the system is that the water of the medium permeates through a semipermeable membrane that allows penetration of water (but not the active agent) to dissolve its content, which is pushed off.

The delivery of an active agent based on osmotic pressure is one of the most promising drug delivery technologies of the last 40 years. Osmotic pumps are utilized from the drug development stage (e.g., determination of pharmacokinetic parameters in animals) to modern drug delivery technologies (e.g., oral controlled-release and implantable systems) (Wright & Stevenson, 1999).

Rose and Nelson (1955), two Australian physiologists, developed a system for use in pharmacological research aiming to deliver drugs to the gut of sheep and cattle. Their osmotic pump (Figure 6.4) was never patented, and consisted of three

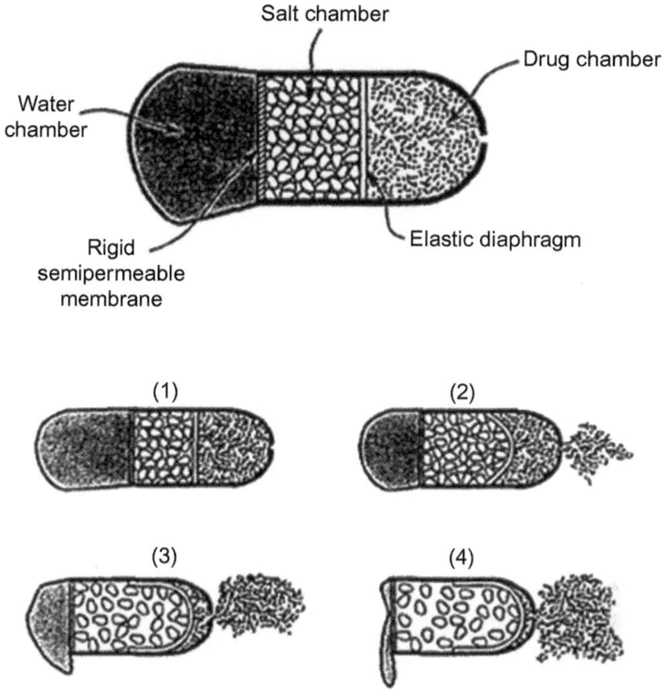

Figure 6.4 Schematic diagrams of the principle of the three-chamber Rose–Nelson osmotic pump. Reprinted from Santus and Baker (1995), with permission from Elsevier.

chambers: a drug chamber (a reservoir containing the delivery orifice), a salt chamber (containing excess solid salt), and a water chamber.

The drug and water chambers were separated by a rigid, semipermeable membrane. The difference in osmotic pressure across the membrane induces water to move from the water chamber into the salt chamber. This phenomenon results in an increase in the volume of the salt chamber, which distends the latex diaphragm separating the salt and drug chambers, thereby pumping the drug out of the system (Santus & Baker, 1995).

From this development, many patents have been proposed, with most of them describing the use of miniature infusion systems strapped to the patient, delivering drugs via an indwelling catheter (Santus & Baker, 1995).

The Rose–Nelson device had limited practical use due to its intricate design, but led to the development of a number of variations. Higuchi and Leeper (1973) proposed osmotic pumps activated by the aqueous media from the site of administration or surrounding environment (Figure 6.5). Their device was composed of two chambers. When the sealed foil was removed from the pouch and the system was placed in an aqueous environment, the pumping action began and the drug was delivered at a constant rate.

Theeuwes and Higuchi (1974) developed osmotic pumps without a separate water chamber, activated by the aqueous media from the site of administration (Figure 6.6).

A milestone in the history of not only osmotic pump development, but also in the history of drug delivery, was the osmotic pump developed by Theeuwes (1975).

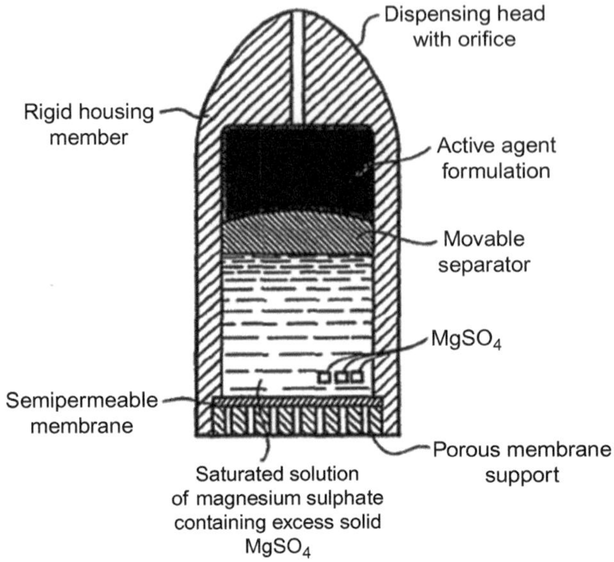

Figure 6.5 Schematic design of the Higuchi–Leeper pump (U.S. Patent 3,760,804). Reprinted from Santus and Baker (1995), with permission from Elsevier.

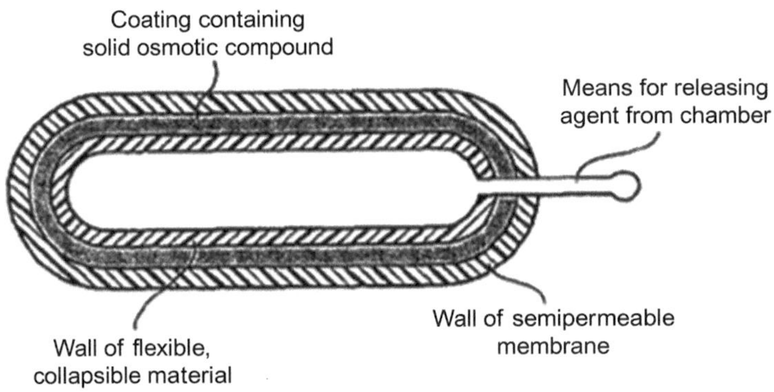

Figure 6.6 Illustration of the Higuchi–Theeuwes pump design described in U.S. Patent 3,995,631.
Reprinted from Santus and Baker (1995), with permission from Elsevier.

It was the first oral osmotic pump (elementary osmotic pump, EOP), in which both the drug and osmotic agent compartment were combined into one, with the possibility of achieving zero-order delivery (Figure 6.7).

Macroscopically analyzing it, the pump is really a tablet with an orifice, representing a simplification of the original. However, the system is composed of a compressed drug, with suitable osmotic pressure into a tablet using a tableting machine. A semipermeable (usually cellulose acetate) membrane coats this tablet, and a small orifice is drilled through the membrane coating. When the device is administered, the aqueous environment and the osmotic pressure of the soluble

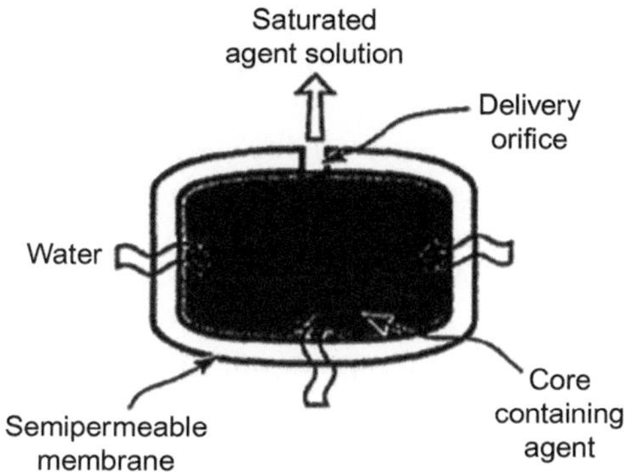

Figure 6.7 Schematic design of the basic concept of the Theeuwes elementary osmotic pump (U.S. Patent 3,845,770).
Reprinted from Santus and Baker (1995), with permission from Elsevier.

active agent inside the tablet draw water through the semipermeable coating, forming a saturated aqueous solution inside the device. The imbibitions of water by the system increase the internal volume. The membrane is non-extensible and the hydrostatic pressure inside the tablet rises slightly. The way to relieve the pressure is to flow the saturated drug solution out of the system through the small hole. The process continues until the entire solid active inside the system has been dissolved and only a solution-filled shell remains. This residual dissolved active agent continues to be released, but at a declining rate, until the osmotic pressures (inside and outside the tablet) equal.

It is important to note that the osmotic pressure produced by the dissolved active agent solution has to be relatively high to overcome the osmotic pressure of the body. However, for active agents displaying solubilities greater than 5–10% (w/w), the system works very well (Santus & Baker, 1995).

6.3.1 Composition

Basically, osmotic pumps possess a core and coatings that are composed of similar components as for matrix controlled systems. The core constitutes the inner structure of the pump and is composed of a therapeutic agent, filler or substrate, viscosity modifier, solubilizer, and lubricant/glidant. The coatings are composed of membrane polymer(s), plasticizer(s), a membrane modifier and a color/opacifier.

Considering that water is the only diffusing species, an agent must be included in the core that is water-soluble enough to generate osmotic pressure and enable the pumping out of the therapeutic agent. In addition, the coating must act as a semipermeable membrane that should be fully coalesced and be free from unintentional pinholes (Collett & Moreton, 2001).

The active agent plus diluents are included in a water-soluble tablet core, which will solubilize (or suspend) the active in the presence of water. The core coating constitutes a semipermeable membrane that will allow water to pass through into the core, establishing an osmotic pressure and forcing drug dispersion to be pumped out of a small hole in the system coating. Therefore, the rate of this pumping of the active agent can be controlled by characteristics of the system like delivery orifice, coating material, and composition.

Together, the active agent, a water-soluble polymer or an inert salt constitute the core. Sometimes, the drug itself constitutes the osmotic component, such as the metoprolol fumarate, whose saturated solution has an osmotic pressure of 32.5 atm (Theeuwes, Swanson, Guittard, Ayer, & Khanna, 1985).

Considering both the clinical and experimental levels, selecting an appropriate dosage pattern of the therapeutic agent for treatment, and understanding the logic for the selection of an active, can be achieved without painstaking empirical investigations by use of osmotic pumps.

6.3.2 Advantages and disadvantages

Osmotic pumps are well-characterized drug delivery systems to be administered by different routes (e.g., *peroral* and subcutaneous). They are well-characterized and understood systems, and have some advantages in relationship to other systems:

- The coating technology is straightforward, in which controlled drug delivery is due to the water (diffusion species), and the modification of the rate of water diffusion is more uncomplicated than for many active agents.
- These systems are suitable for a wide range of therapeutic agents, and they typically give a zero-order release profile after an initial lag.
- The delivery rate of active(s) is highly predictable and programmable.
- The release mechanism is not dependent on the active agent.
- Higher release rates can be achieved than with conventional diffusion-based drug delivery systems.
- A high degree of *in vitro/in vivo* correlation can be obtained.
- Drug release from the oral osmotic pumps is minimally affected by food presence.

On the other hand, osmotic pumps have some disadvantages. First, the orifice size is critical and the laser drilling is very important. The membrane must be fully controlled and therefore the coating process must be well controlled, avoiding the risk of film defects, which could result in dose dumping. Thus, the film droplets or particles must be induced to coalesce into a film with consistent properties (Collett & Moreton, 2001).

One advantage of this strategy is that the delivery of the active agent is not affected by pH (e.g., gastrointestinal acidity or alkalinity) nor stirring rate (e.g., fed conditions or gastrointestinal motility). On the other hand, the delay in active delivery is affected by a slowly solubilized coated layer between the active drug core and the outer semipermeable membrane. Changing the surface area, thickness, or composition of the membrane and/or diameter of the drug release hole can lead to alterations in drug delivery.

Moreover, the components of the system are biologically inert and remain intact during the time of action (inside the subcutaneous tissue or in the gastrointestinal tract). In the case of subcutaneous administration, the system needs to be removed after therapy. However, when the system is administered through *peroral* route, it is eliminated in the feces as an insoluble shell.

6.3.3 Drug delivery rate

The drug delivery of the active agent from osmotic pumps across the semipermeable membrane is controlled by water movement.

Therefore, for one solution-type osmotic pressure-activated system, the intrinsic rate of active agent delivery is defined by Equation (6.5):

$$\frac{Q}{t} = \frac{P_w A_m}{h_m}(p_s - p_e) \tag{6.5}$$

where Q/t is the intrinsic rate of drug delivery, P_w is the water permeability in the semipermeable membrane, A_m is the effective surface area of the semipermeable membrane, h_m is the thickness of the semipermeable membrane, and $p_s\_p_e$ is the differential osmotic pressure between the system (with an osmotic pressure p_s) and the environment (with an osmotic pressure p_e).

For a solid-type osmotic pump system, the intrinsic rate of delivery is defined considering the drug concentration (aqueous solubility of the active agent) in the solid reservoir (S_d), according to Equation (6.6):

$$\frac{Q}{t} = \frac{P_w A_m}{h_m}(p_s - p_e)S_d \tag{6.6}$$

Considering that the system possesses constant internal volume, it will deliver in any time interval a volume of saturated solution equal to the volume of solvent uptake. Moreover, the drug delivery rate will be constant during the presence of an excess of active agent inside the device. When the concentration falls below saturation, the delivery rate will decline parabolically toward zero.

The size of the orifice is a key factor in order to produce a proper drug delivery. In this sense, the orifice size must be optimized in order to minimize the contribution to the delivery rate made by solute diffusion through the orifice and the hydrostatic pressure inside the system, which would affect the zero-order release rate. When the orifice is too small, it will depress the delivery rate below that of the desired constant delivery.

Depending on the administration route, some problems may occur. This is the case when the pump is administered by *peroral* route, in which control of the transit of the system down the gastrointestinal tract may vary considerably with time (e.g., when the system is designed to release the therapeutic agent over a period of 10h, but the total transit time in the gut is 5h). In this case, the drug bioavailability is reduced (Florence & Attwood, 2006).

6.3.4 Classification

Osmotic pumps constitute a versatile strategy that is used as an important tool to determine various pharmacokinetic parameters and the pharmacodynamic response of active agents. In addition, they can be orally administered and implanted for controlled, targeted or pulsed delivery of active agents during the desired period of time.

During the last 30 years, the osmotic patent field has been dominated by Alza Corporation, and several oral osmotic pumps systems have been successfully marketed for therapeutic uses. Therefore, osmotic pumps can be classified as osmotic pumps for experimental research and osmotic pumps for drug delivery (Verma, Arora, & Garg, 2004).

Alzet® osmotic pump, as indicated by its name, was a system developed by Alza Corporation for experimental research. The osmotic delivery device is based on Higuchi–Theeuwes pump design (Figure 6.6). It became available for investigational use in the mid-1970s, and researchers used it routinely in the screening of active agents, animal toxicology and pharmacology, and initial clinical testing, creating new therapeutic options.

In cross section (Figure 6.8), they are composed of a drug core (reservoir), the osmotic agent, and the semipermeable membrane (rate controller). In addition, a flow moderator

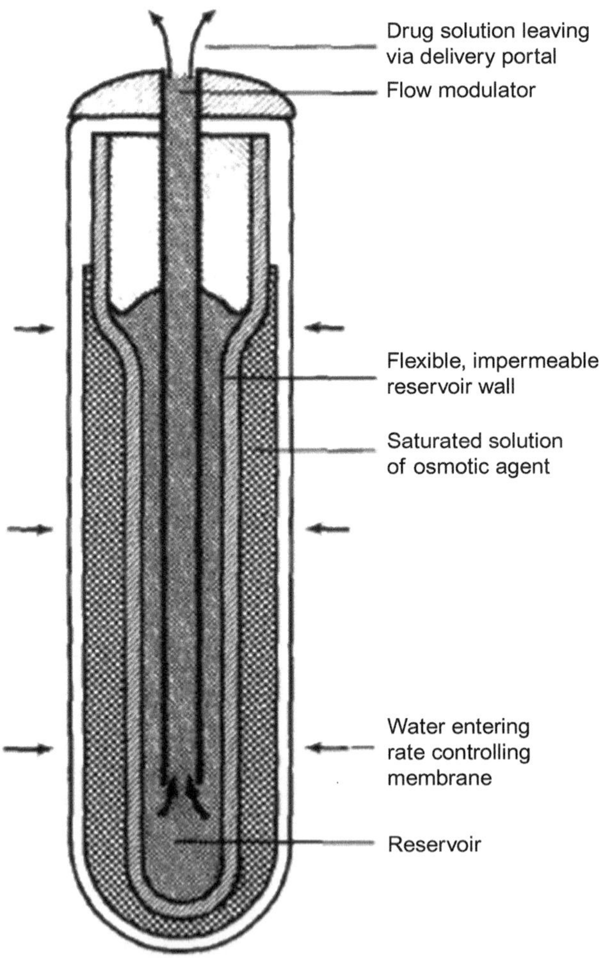

Figure 6.8 Cross-section of ALZET® osmotic pump showing the principle of the functioning. Reprinted from Amkraut, Eckenhoff, and Nichols (1990), with permission from Elsevier.

is inserted into the body of the osmotic pump after filling. It is a type of implantable or insertable system in which the active agent is a solution or suspension contained in a cylindrical reservoir formed from a synthetic, collapsible, impermeable elastomer wall (e.g., polyester) that is open to the external environment via a single orifice.

Many active agents have been delivered by the Alzet® system (e.g., amino acids, anesthetics, antibiotics, antibodies, anticancer agents, anticoagulants, anti-epileptics, antigens, antihypertensives, anti-parasitic agents, anti-Parkinson agents, ascitic fluid, catecholamines, chelators, cholinergics, central nervous system acting agents, enzyme inhibitors, and gastrointestinal modulators).

The wall of the active agent compartment (semipermeable membrane) is chemically inert to the most active-vehicle combinations (Fara & Ray, 1988): purified water or any aqueous solution, isotonic NaCl or other salt solution, aqueous solution of dextrose (5%), ethanol (up to 10% in water), polyethylene glycol 300 (neat or in water), aqueous solution of Tween (2%), dimethyl sulfoxide (50% in water or 50% in ethanol 10%), propylene glycol (neat or in water), glycerin, dilute acids (with pH greater than 1.8), dilute bases (with pH less than 14), rat serum, bacteriostatic culture media (1% benzyl alcohol as bacteriostatic), and artificial cerebrospinal fluid. In addition, it needs to be impermeable, avoiding the crossing of any material between the reservoir and the surrounding osmotic agent. This last one is located outside the reservoir's semipermeable membrane, forming a thin cylinder.

The core (reservoir) of the pump can be filled with the appropriated dispersion containing the active agent though the delivery hole, using a special filling tube attached to a syringe, avoiding the introduction of air. Following that, a flow moderator is inserted though the orifice into the body of the pump, with the aim to minimize the diffusion of the active out of the reservoir. This ensures that the osmotic pressure will control the release, providing a convenient attachment point to the catheter. The absence of air inside the reservoir can be determined by calculating the weight difference between the unfilled pump and the filled pump with the flow moderator in place.

The external surface is coated with a layer of osmotically active salt (e.g., sodium chloride). The reservoir compartment is completely sealed inside a rigid compartment walled with a semipermeable membrane.

The temperature exerts influence on the membrane permeability and on osmotic pressure of the osmotic agent (Figure 6.9).

Alzet® mini-osmotic pumps have demonstrated that the delivery profile is maintained at different temperatures. Only the magnitude and the duration of delivery change. By varying the concentration of the active agent in the dispersion, the amounts delivered can be varied at a constant temperature (Amkraut et al., 1990). They have capsular shape and are available in a variety of sizes, and it is possible to relate the delivery rates with the reservoir capacities (Table 6.4).

Strict specifications are used in the manufacture of these pumps, and studies have shown that the rates of volume delivery *in vitro* and *in vivo* (as an implant) are within 5% of each other (Amkraut et al., 1990).

The Osmet® modules use the Alzet® technology and are used as research tools in human pharmacological studies. They can be developed for rectal, vaginal, or oral administration.

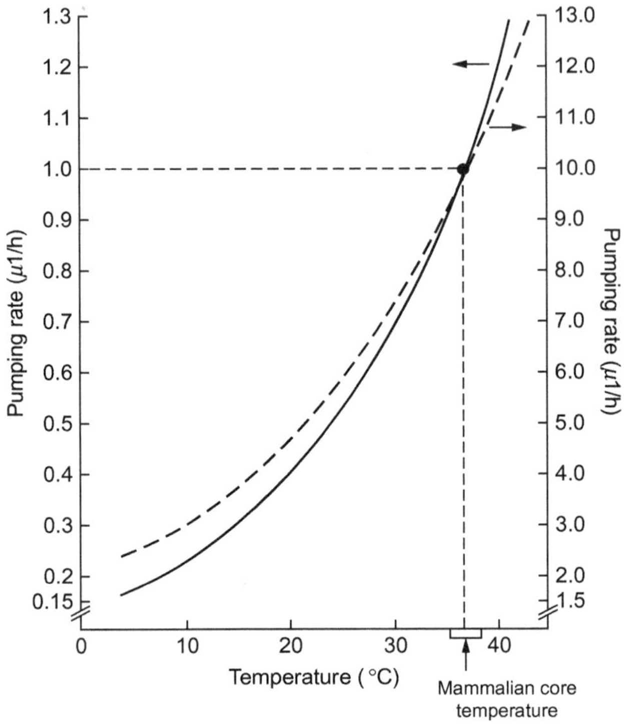

Figure 6.9 Graphical illustration displaying the temperature dependence of the delivery rate of ALZET® mini-osmotic pumps. The solid line represents pump volume; the dotted line represents 2-ml pump volume.
Reprinted from Amkraut et al. (1990), with permission from Elsevier.

Table 6.4 Different types, reservoir capacities, and delivery rates of ALZET® osmotic pumps

Type	Capacity	Delivery rate (µl/h)	Period of delivery
1003D	100 µl	1.0	3 days
1007D	100 µl	0.5	1 week
1002	100 µl	0.25	2 weeks
1004	100 µl	0.11	4 weeks
2001	200 µl	1.0	1 week
2001D	200 µl	8.0	1 day
2002	200 µl	0.5	2 weeks
2004	200 µl	0.25	4 weeks
2006	200 µl	0.15	6 weeks
2ML1	2 ml	10	1 week
2ML2	2 ml	5.0	2 weeks
2ML4	2 ml	2.5	4 weeks

They can be utilized in clinical studies to deliver a variety of active agents. The definition of the relationship among rate, extent, and time of absorption to the design and efficient development of rate-controlled, duration-specified oral systems of a particular therapeutic agent can be accomplished by use of Osmet® osmotic pumps.

It is also possible to design osmotic systems for colon-targeted drug delivery (OSMET®-CT). The enteric coating present in this type of osmotic pump enables the delay of the start of drug delivery for 2–4h after passing from the stomach. Actually, the Osmet® pumps are designed to be relevant to standard gastrointestinal transit times (8–24h). The drug delivery rates are independent of pH, and the delivery rate profiles are similar to those from the Alzet®.

Other examples of osmotic pumps are the elementary and push-pull OROS® delivery system. As previously discussed, the first osmotic pumps were developed in the 1950s. It was very difficult to use them in humans because they had a very complicated design. However, with the development of oral osmotic pumps, osmotic drug delivery technology constituted the major method to achieve controlled drug delivery in humans.

Considering the route of administration in humans, osmotic pumps can be classified into two categories: oral osmotic pumps and implantable pumps.

Since the design of the first oral osmotic pump developed by Alza in 1975, a large number of systems have been developed based on this strategy, which is the OROS® system.

The OROS® system is characterized to possess a solid core, containing an active agent or in the absence of an osmotic agent, coated with a semipermeable membrane.

The simplest oral osmotic system is the single-compartment EOP, which is suitable for active agents that demonstrate an intermediate solubility range. On the other hand, for extremely soluble or insoluble active agents, the push-pull OROS® system (Figure 6.10) is the most useful. In this type of osmotic pump, the water enters the two layers at essentially the same rate. Dispersion of the active agent is formed in the reservoir by the controlled ingress of water, which is pushed out through a hole (0.4mm diameter) present in the semipermeable membrane. Only a small amount of water is

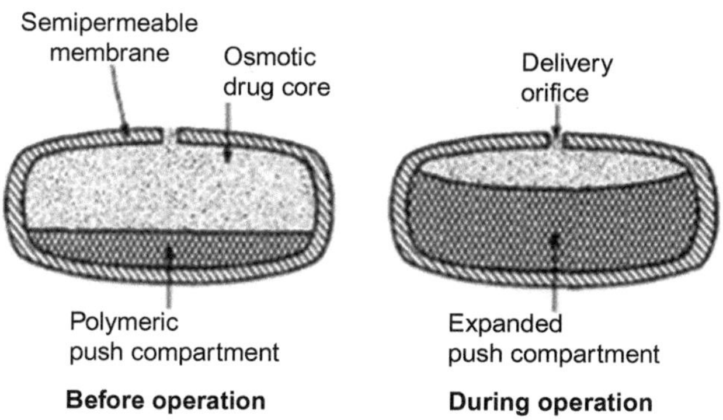

Figure 6.10 Push-pull OROS® osmotic pump.
Reprinted from Amkraut et al. (1990), with permission from Elsevier.

drawn into the system each hour and is dependent on the osmotic gradient between the contents of the two-layer core and dissolution medium.

Both types of osmotic pumps (elementary and push-pull) provide predictable release of the active agent *in vivo* based on *in vitro* tests. Moreover, the drug release rate is independent of pH and gastrointestinal tract motility patterns (Amkraut et al., 1990).

The technology of the OROS® system began to be used with the first product, Osmosin® (controlled-release indomethacin), which was withdrawn a year after launching because of side effects (Verma et al., 2004). After that, a great number of other products launched to delivery: vitamin C (Acu System® C by Alza), phenylpropanolamine (Acutrim® by Alza), fexofenadine hydrochloride and pseudoephedrine hydrochloride (Allegra D by Aventis/Osmotica), lovastatin (Altoprev® by Andrx; Cardura® XL by Alza/Pfizer/Andrx), and nifedipine (Procardia XL® by ALZA/Pfizer in the United States and Adalat CR® by Bayer in Europe). The osmotic tablet of nifedipine contains the semipermeable cellulose acetate coating, the swellable hydrogel layer of polyoxyethylene glycol and hydroxypropylmethylcellulose, and the drug chamber containing nifedipine in hydroxypropylmethylcellulose and polyethylene glycol. Nowadays, there is a variety of products in the market (Verma et al., 2004; Ghosh & Ghosh, 2011).

When the osmotic pump is designed to be subcutaneously implanted at the desired site of the body, the water content in the tissue fluid will penetrate through the semipermeable membrane at a controlled rate and dissolve the osmotically active agent. Osmotic pressure is produced in the narrow spacing between the flexible reservoir wall and the rigid semipermeable compartment. The reservoir is gradually reduced in volume and the solution of the active agent is forced to exit through the flow moderator at a controlled rate (Theeuwes, 1975; Theeuwes & Yum, 1976).

Therefore, by varying the concentration of the active agent in the solution, different doses of active can be released at a constant rate as long as the osmotic gradient remains constant.

DUROS®, a long-term zero-order drug delivery system designed by Alza, was the first implantable osmotic pump to be used in humans (Figure 6.11). It is a small pump (4 mm diameter × 45 mm length) composed of a titanium alloy cylinder and capped on one side by a rate-controlling semipermeable membrane (polyurethane polymer), and

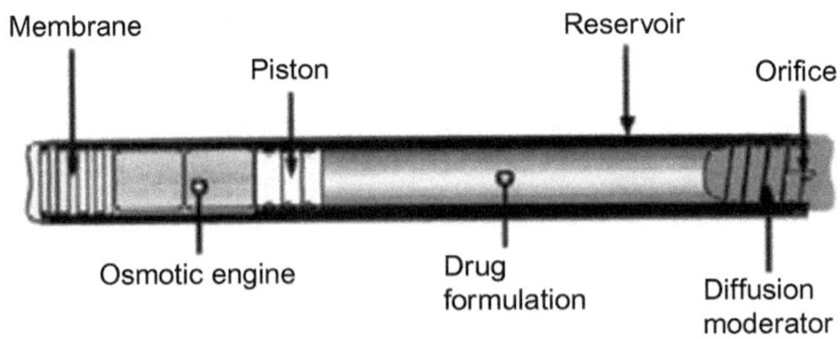

Figure 6.11 Illustration showing the cross-sectional view of DUROS® implantable osmotic system.
Reprinted from Wright et al. (2001), with permission from Elsevier.

on the other by a plug with a hole for active agent delivery. Only water can permeate the membrane. A piston made of elastomeric materials is located on the side of the pump, separating the osmotic site (normally containing sodium chloride) from the active agent reservoir. The extracellular body fluids activate the osmotic engine and the piston displaces, resulting in delivery of the active agent.

Viadur® (leuprolide acetate) was the first therapeutic application of the DUROS® osmotic pump. The implantation system was designed for the treatment of prostate cancer, displaying the property to provide the delivery of leuprolide acetate continuously at a rate of 125 μg/day for 1 year (Wright et al., 2001).

Moreover, this technology was extended to the controlled delivery of active agents in the rectum by zero-order kinetics, avoiding the hepatic first-pass metabolism.

6.4 Capsules

Capsules are a type of solid-dosage form in which the active agent is contained inside a reservoir. Normally, they are composed of gelatin, starch or cellulose derivatives (e.g., hydroxypropylmethylcellulose or hypromellose), rendering two types: hard and soft capsules.

Capsules can be different in design and properties. Soft capsules are more flexible and are composed of just one-piece capsule shell (Figure 6.12), whereas the hard ones are less flexible and are composed of two pieces (Figure 6.13).

This dosage form enables the administration of a wide range of preparation types. These preparations may be composed of several components, which have the function to facilitate the manufacture or to control the therapeutic performance of the dosage form.

The active agent plus the other components of formulation are added within the interior of the capsule. In this context, granulates, powders, mini-tablets, non-aqueous liquids and gels, and semi-solids may be filled into hard capsules.

Solids are accommodated in standard two-piece capsules (Figure 6.13), whereas liquids and semi-solids can be contained using capsules designed for liquid fillings

Figure 6.12 Soft gelatin capsules.

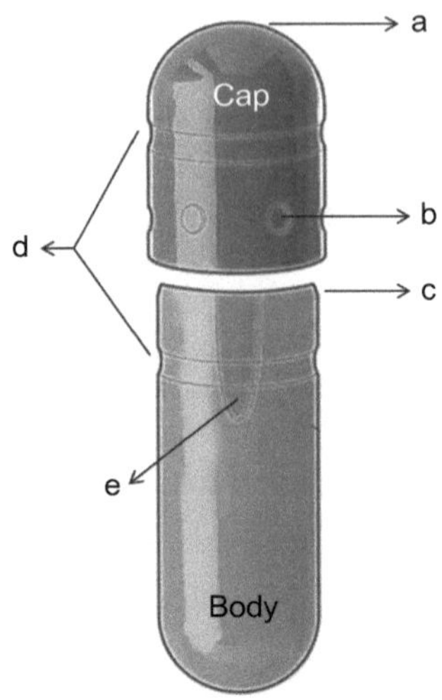

Figure 6.13 Hard capsules (ConiSnap®): (a) rounded hemispherical ends; (b) elongated dimples to maintain the precise round diameter; (c) tapered rim of the body engages easily with the cap for problem-free closure; (d) closely matched locking rings provide full-circumference leak-free closure; (e) two aerodynamic air events allow air to escape from the cap.

(e.g., Licaps capsules) and readily sealed by microspray technology (e.g., LEMS sealing) or band sealed with gelatin (Stegemann, 2011).

Soft capsules, normally composed of gelatin, are filled with non-aqueous liquids containing the active agent dispersed within this carrier.

Gelatin is a suitable material to produce capsules because it possesses excellent physicochemical and biological properties. It is nontoxic and soluble in biological fluids, it undergoes a sol-gel transition at relatively low temperatures, and it displays excellent mechanical properties.

For the manufacture of capsules, specialized equipment is necessary. Capsules containing liquid fills have potential problems of stability. Moreover, the homogeneity of fill weight and content are also associated with this dosage form. However, capsules may increase the oral bioavailability of poorly soluble active agents, enable a convenient oral administration of liquids to patients as a unit dosage form, and improve the stability of the active agents.

Gelatin is a mixture of proteins from animal collagen (e.g., from animal skins, sinews and/or bovine bones) by either partial acid or partial alkaline hydrolysis. Two types of gelatin may be obtained from these processes: type A and type B. These two types of gelatin differ in their isoelectric point and solubilities, which are in function of pH.

Commercially, it is possible to purchase the two types of gelatin as well as mixtures of them.

6.4.1 Hard capsules

Hard gelatin capsules are designed to have two pieces (cap and body). Their manufacture is performed using a dip-coating method by which the gelatin solution receives the other constituents of the capsule (e.g., colorants, opacifying agent, wetting agents, and lubricants), the viscosity of the solution is reduced, and the metal pins (molds) are dipped into a pan containing the final gelatin solution. The cap and the body of capsules are produced using different molds, and their final water content should be of 13–16% w/w. This is important because the water acts as a plasticizer for gelatin, ensuring suitable mechanical properties of the capsule (avoiding cracking or permanent deformation during manufacture, handling or storage). In this sense, controlling the environmental conditions of capsule storage is very important.

A wide range of capsule sizes is available, enabling the accommodation of fill volumes between 0.20 and 0.67 ml. During the filling of hard capsules, the preparation is added in the body, using a range of different mechanical techniques, and the cap is pushed into place. Body and cap are joined, overlapping the body with the cap. The presence of closely matched locking rings provides full-circumference leak-free closure, and the separation of body and cap does not normally occur under normal storage conditions or in clinical use.

6.4.2 Soft gelatin capsules

Soft gelatin capsules are composed of gelatin, plasticizing agents, water, and miscellaneous excipients. They display different mechanical properties as a consequence of the addition of a plasticizer (e.g., glycerol, sorbitol, or other polyhydric alcohol), which results in a more flexible structure.

They enable the possibility to develop liquid-based formulations that provide a greater C_{max} than tablet formulations. Oblong, round, tubular, and oval are the main shapes of soft gelatin capsules.

6.4.3 Capsules as drug delivery system

Gelatin capsules were one of the first drug delivery systems for pharmaceutical preparations. Nowadays, capsules constitute a type of dosage form that can modify and control drug delivery. Multiparticulate preparations coated to control release and a release-controlling matrix with an active agent can be used to fill the capsules.

According to the composition of the capsule wall, it is possible to enable immediate or modified release of the active agent. Therefore, soft gelatin capsules are used for immediate-release liquid products and for therapeutic agents requiring solubility-enhancing formulations (e.g., complex emulsion systems or formulations containing hydrophobic solvents). Moreover, immediate- and modified-release products can be achieved using hard capsules, which start to open their interface within about 2 min in aqueous media at 37 °C.

When the liquid penetrates, the capsule wall wets and swells, and the system gelling affects the delivery of the formulation contained in its interior. The swelling and gelling may be dependent on pH and ionic strength of the dissolution medium. Thus, it is possible to use the cross-linking and coating to modulate the drug delivery.

When the capsules are suitably coated, they can modulate the delivery of the active agent in the desired site of the gastrointestinal tract (e.g., stomach or intestine). In this sense, the pharmaceutical industries have utilized hard capsules to modify the release of omeprazole, esomaprazole, tamsulosin, venlafaxin, and dexlansoprazole (Olsen & Hitzeman, 2009).

Compared to tablets, the capsule dosage form has some advantages, like the avoiding of rupture of membranes caused by compression and the non-compression of capsule content. Moreover, it is possible to encapsulate, without compression or stress, coated mini- or microparticulates that control drug release, matrix modified-release particles prepared by extrusion or granulation with release-modifying polymers, mini-tablets providing different rates or modes of release, mixtures of particles with differing release rates, and mixtures of immediate and controlled-release units (Stegemann, 2011).

The coating of capsules can prevent the gelatin dissolution at stomach pH (low) and, therefore, it can be used to protect the acid-unstable active agent.

Delayed delivery can be also obtained by the use of coated hard capsules. The coating can produce protection for the first part of the gastrointestinal tract and deliver the active agent along the second part (e.g., colon and rectum). Another option is to provide a mixture of immediate delivery (uncoated) and delayed delivery to prolong plasma levels.

In these cases, pH or enzymatically sensitive coatings are utilized. The most used polymers are polymethacrylates (e.g., Eudragit®), cellulose-based polymers such as cellulose acetate phthalate (e.g., Aquateric®) and polyvinyl derivatives such as polyvinyl acetate phthalate (e.g., Coateric®). Plasticizers such as triacetin, triethyl citrate, diethyl phthalate, silicon oil, and acetyltriethyl citrate may also be present. Materials such as talc, magnesium stearate, and Aerosil® may be included as anti-adhesion agents. Titanium dioxide or iron oxides can provide color (Stegemann, 2011).

6.5 Micro- and nanoparticles

Over the past 40 years, the encapsulation of substances in small particles has become a well-established technique of compartmentalization, allowing active agents not to come into contact with the external environment until the moment their action is needed.

The microencapsulation process enables solids, liquids or even gases to be retained in microscopic structures by the formation of a matrix or one or more thin coatings of a material. This strategy was first used in the late 1930s by the National Cash Register Company (NCR®) in the production of carbonless copying paper (no carbon required). In this process, two compounds separated by a clay film, when together, react with each other yielding color, which is caused by the mixture of the film breaking at the time of writing. In the 1950s, NCR® developed a reproduction paper and ribbons containing dyes microencapsulated in gelatin and other polymeric materials (e.g., polyvinyl alcohol, ethylcellulose, and polyvinyl chloride). With the impact by a typewriter key or the pressure of a pencil or pen, the dyes were released.

In the 1970s, microencapsulation was considered less science than art, because most of the research was developed in pharmaceutical companies and very little information was discussed in scientific meetings. The success of this strategy led scientists to apply microencapsulation technology to other areas; today, it is currently used in pharmaceuticals, printing, agriculture, the food industry, household products, and industry textiles, among others (Benita, 2006; Allen et al., 2011).

In the pharmaceutical field, microencapsulation has been extensively studied, particularly for obtaining solid form oils, flavor and odor control, drug protection against moisture, heat and oxidation, alteration of solubility and decreased volatilization, and for preventing incompatibilities, making toxic materials more easily manipulated, improving flow characteristics and enabling controlled drug release and drug vectorization, among others.

Nowadays, micro- and nanoencapsulation is an important field for the development of new formulations, because it allows, in a rational and effective way, increased therapeutic effectiveness of substances that are used in the treatment of a great variety of diseases. Furthermore, it makes possible the use of potentially toxic drugs, as many drugs are.

The "microparticle" and "nanoparticle" terms are generic and are used according to the particle size to which they refer. Particle sizes less than 1 µm are considered particles with nanometric size, whereas larger particles are called microparticles. Particles larger than 1000 µm are classified as macroparticles, although there are authors who see them as microparticles as well. In addition, particles showing a size up to 250–500 nm are considered nanoparticles.

The micro/nanoparticle term is broad and refers to two different structures: micro/nanospheres and micro/nanocapsules (Figure 6.14). They may have a regular (e.g., spherical, tubular, and oval) or irregular shape. Although the terms tend to be used interchangeably, we retain the distinction here. Micro/nanocapsules are generally spherical particles if they are carrying a liquid or gas, or another format if they are carrying semisolid or solid particles, which are a system-type reservoir. Micro/nanospheres are those systems in which the drug is homogeneously dispersed within the polymer matrix or wax, forming a monolithic system where it is not possible to identify a distinct nucleus.

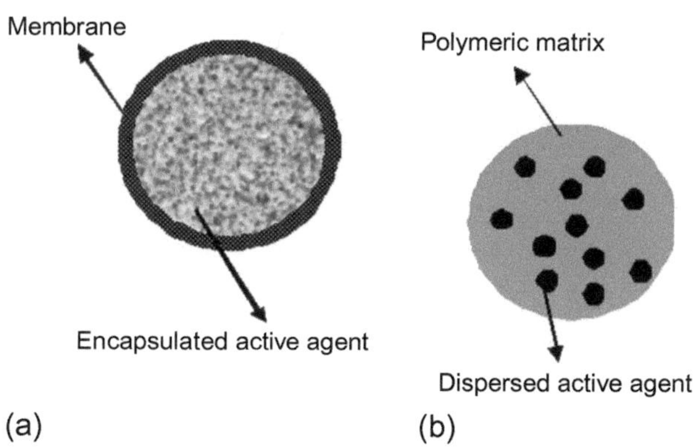

Figure 6.14 Structure of systems obtained by microencapsulation: (a) microcapsule or reservoir; (b) microsphere or matrix.

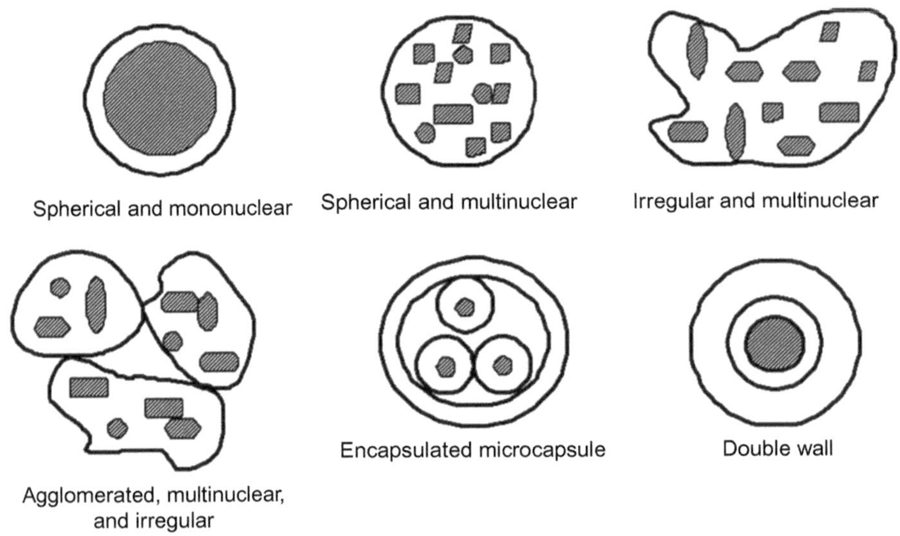

Figure 6.15 Some types of microparticulate systems.

The first microparticles were obtained with sizes ranging from 5 to 2000 μm. However, since 1980, a second generation of microparticulate systems of various types (Figure 6.15) and much smaller dimensions has been developed. These systems include nanoparticles and microparticles. A third generation of systems has been studied, with objectives to target drugs contained in nanoparticles, liposomes, niosomes, and nanoemulsions to specific sites, favoring the drug action in the body or desired tissue (Benita, 2006).

Especially in the field of drug delivery, nanotechnology is allowing real progress to achieve temporal and spatial site-specific delivery. The pharmaceutical sciences have observed a great increase in the nanotechnology and drug delivery systems market. The number of patents and products in this field is increasing significantly as well (e.g., for cancer treatment, brain disorders, vaccines, and diagnostics).

Nanoparticulate drug delivery systems have been described using several terminologies, as previously discussed. In addition, polymers and lipids are the main types of materials used as carriers for the active agent, resulting in delivery systems showing particle size distribution from a few nanometers to a few hundred nanometers.

Therefore, it is possible to classify the types of carriers according to their composition:

(1) Polymeric systems (dendrimers, polymer micelles, niosomes, microparticles, nanoparticles, nanogels, polymer-drug nanoconjugates, chitosan polymers, methacrylate polymers, bioadhesive systems, thermoresponsive systems).
(2) Lipid systems (solid lipid nanoparticles, lipid nanostructured systems, cubosomes, liposomes, polymerosomes, immunoliposomes, microemulsions, nanoemulsions, liquid crystals).

(3) Proteins/peptides (microparticles, nanoparticles, peptide nanotubes, fusion proteins, immunotoxins).
(4) Metal structures (metal colloids, carbon nanotubes, fullerenes, gold nanoparticles, gold nanoshells, silicone nanoparticles, magnetic colloids).

In the following section, the discussion will focus on nano- and microparticles.

6.5.1 Properties and applications

The microencapsulation strategy has many properties and applications. It can modify the color, shape, volume, apparent density, reactivity, durability, pressure sensitivity, heat sensitivity, and photosensitivity of the encapsulated substance. The microparticle can preserve a substance in a finely divided state and release it as required.

Microparticulate systems generally consist of polymers. However, they may also be obtained using proteins, lipids, or waxes. Drugs, when encapsulated within the polymeric particles, are not readily available to the biological system because the polymer must swell or be degraded so that the drug can be released, or the drug has to dissolve and/or diffuse within the polymeric particles. In any event, the release of the active agent into the physiological environment is extended for a much longer time compared to the unencapsulated agent. This feature is one of the main factors in the research and development of microparticle systems, because it can reduce the number of doses and the fluctuations of plasma levels of drugs administered, allowing a more effective therapy.

The encapsulation of liquids effectively converts them to a powder. Thus, microencapsulation promotes easy handling of liquid reagents, starting the reactions only after the system is broken. The density of a substance can be enhanced by encapsulation and can be decreased by incorporation of air into the capsule. The storage time of volatile compounds can be increased by microencapsulation. Microparticles can protect a substance from the effects of ultraviolet (UV) rays, moisture, and oxygen. Chemical reactions between two substances can be avoided by physical separation using microencapsulation. Finely divided powders can be also microencapsulated to reduce agglomeration problems (Brandau, 2002).

Micro/nanoparticles have many useful functions and have been employed in many fields of technology, often linked with applications in which the contents are released under controlled conditions. A wide variety of different products is micro- or nanoencapsulated for applications:

(1) Food industry: Many foods have been encapsulated in gelatin and vegetable gums. Food additives are encapsulated to increase storage time by reducing oxidative degradation and volatilization. Examples include flavors and aromas.
(2) Encapsulated pigments (e.g., cosmetics) are prepared using acrylic copolymers as film formers.
(3) Paper: The main application is in the copy papermaking.
(4) Pest control and weeds: There are many applications of microcapsules in agriculture, increasing the release of the active ingredient (pesticide) in the required target. In many cases, the microparticles have been made for specific products for particular uses. The properties of microparticles are associated with the biological behavior of the living organism that must be controlled. Microcapsules have been designed for sustained and controlled release of biologically active materials such as pesticides, to promote longer effectiveness.

(5) Articles of textile industry: The microencapsulation strategy is increasingly present in tissues in the control of perspiration odor, mainly in socks and shirts in the application of durable fragrances and softeners.
(6) Cosmetic products: Microencapsulation is used in the production of encapsulated perfume and makeup removers, for example.
(7) Pharmaceutical field: There are many applications, such as acetylsalicylic acid in ethyl cellulose and other cellulose esters to reduce gastric irritation. The manufacture of microparticles resistant to stomach acid pH, which releases the drug in the high gut pH, is another example. Many microencapsulated agents have controlled release, with specific rates. Biologically active substances (e.g., deoxyribonucleic acid (DNA), entire cell or group of cells) can have their performance improved and/or their shelf life enhanced by the use of encapsulation.
(8) Other uses of loaded microcapsules: Essences on paper have been widely used in marketing and promotion of products (scratch and sniff). Microencapsulation can also be used to encapsulate cells and enzymes and to make artificial cells. Other microencapsulated products include dental amalgam containing mercury encapsulated, photographic materials and thermosensitive liquid crystals, microencapsulated catalyst reagents and others.

Some examples of micro/nanoencapsulation of active agents that are already available are analgesics (aspirin; indomethacin); steroids (progesterone, prednisone); vitamins (vitamin A and K, the B calcium pantothenate, vitamin C); sedatives and hypnotics (barbiturates); antibiotics (sulfamerazin; sulfadiazine, sulfamethoxazole, ampicillin, erythromycin); chemotherapeutic agents (methylglyoxal, mitomycin, 5-fluorouracil, doxorubicin); prostaglandins and radiopharmaceuticals (Benita, 2006).

As drug-releasing systems, microparticles are suitable for oral, parenteral, or intravenous administration. Due to their small size, the majority of active agents can be widely distributed via the gastrointestinal tract, thereby greatly increasing absorption.

In addition, because of their particular characteristics, nanoparticles can be administered by parenteral route and can gain access to tissues beyond the capillary blood supply. They can diffuse in tissues more easily than larger particles.

Therefore, micro/nanoencapsulation is a strategy to entrap an active agent (or a system) in a dispersed material with the aim of structuration, functionalization, protection, immobilization, and controlled release.

6.5.2 Preparation

The development of a microparticulate system must take into account the total system: the particle itself, the release mechanism, and the components. Each of these parameters needs to be optimized in order to obtain a satisfactory product. Therefore, many microencapsulation methods exist, according to the characteristics of the material to be encapsulated, the solubility of the wall-forming material, the particle size, wall thickness and permeability, release rate, and physical properties of substances.

In the choice of the microencapsulation process, various physical properties such as solubility and capacity of the core to be surrounded by the wall material should be considered. The core may not be soluble in the solvent of the wall-forming polymer and it must not be substantially immiscible in the liquid core. In the spray coating, a water-soluble solid can be coated with a water-soluble polymer solution, due to the fast evaporation of water during formation of the microparticle.

The elasticity of the wall is determined by the nature of the material (e.g., polymer or blends), the thickness and the size of microparticles, because its permeability determines how long the core content may be released or retained in a predetermined rate, such as in controlled-release applications. The adhesive characteristics of wall-forming materials are markedly affected by the temperature and concentration, depending on the physical properties of the polymeric wall material (melting point, glass transition temperature, crystalline degree, and rate of degradation).

Microencapsulation materials may be of natural, semi-synthetic and synthetic origin, and are selected according to the physical properties of the core and the intended application:

(1) Natural materials: acacia gum (gum arabic), agar, agarose, maltodextrin, sodium alginate, calcium alginate, dextran, fats and fatty acids, cetyl alcohol, milk solids, gelatin, gluten, albumin, starch, caseinates, stearin, sucrose, and waxes (e.g., beeswax, carnauba wax, and spermaceti).

(2) Semi-synthetic materials: cellulose acetate; cellulose acetate butyrate; cellulose acetate phthalate; cellulose nitrate; ethylcellulose; hydroxypropylcellulose; hydroxypropylmethylcellulose phthalate; methylcellulose; sodium carboxymethylcellulose; myristyl alcohol; glycerol mono- or dipalmitate; oil hydrogenated castor mono-, di- or tristearate and glycerol 12-hydroxystearyl alcohol.

(3) Synthetic materials: acrylic polymers and copolymers [e.g., polyacrylamide, cyanoacrylate polyalkyl and poly(ethylenevinyl acetate)], aluminum monostearate, carboxyvinyl polymers (Carbopol®), polyamides, poly(methyl vinyl ether maleic anhydride), polycarbonates, polyterephthalamide, polyvinyl acetate phthalate, polyarylsulphones, poly(methyl methacrylate), polyvinylpyrrolidone, polydimethylsiloxane, polyoxyethylene, polyester, polyglycolic acid and copolymers, polyglutamic acid, polylysine, polystyrene, polyimides, and polyvinyl alcohol.

The physicochemical properties of the materials to be used in the preparation of microparticulate systems are of great importance when applying the controlled release, prolonged and/or drug targeting, because the pharmacokinetics of the encapsulated drug becomes dependent on the system in which it is contained.

Microencapsulation techniques have been proposed with many variations in details, depending on the material to be encapsulated and polymer solubility, particle size, thickness, and wall permeability, the type and rate of release required, physical properties and economic feasibility production. Many of these methods are based exclusively on physical phenomena. Some utilize chemical reactions such as polymerization or polycondensation. Others combine physical and chemical phenomena. Nowadays, there is a huge and increasing number of encapsulation processes, new patents and advanced developments. Therefore, a systemic nomenclature for encapsulation classification can be obtained.

Many microencapsulation systems employ a combination of various processes, and sometimes it is difficult to classify the methods. In the pharmaceutical area, the main technologies can be classified into three groups. The first is the group of technologies that use physicochemical methods:

i. Coacervation (phase separation)
ii. Methods using emulsification
iii. Methods using supercritical fluid
iv. Thermal gelation

Chemical methods compose the second group:

i. *In situ* and interfacial polycondensation
ii. Gelation
iii. Polymerization

The third group is composed by technologies that involve mechanical (physical) methods:

i. Solvent evaporation
ii. Spray drying
iii. Airflow or fluidized bed
iv. Droplet freezing
v. Droplet gelation
vi. Extrusion
vii. Centrifugation

The coacervation (phase separation) technique is based on the desolvation phenomenon of macromolecules, and it leads to phase separation in initially homogeneous colloidal solutions. All factors that modify the colloid solvating, or that interfere with the properties of the solvent in charge of macromolecules, induce coacervation.

This employs a very common phenomenon of polymer–polymer incompatibility, in which the polymer that has become the capsule wall material is dispersed, and to this dispersion another polymer is added, inducing the phase. The two polymers are incompatible and form two phases. The wall-forming material is induced to separate as a viscous liquid phase by different methods, such as by varying the temperature, pH, electrolyte concentration, addition of a nonsolvent or by adding a second material with high solubility in the polymer core or opposite charge. Therefore, reducing the solubility of the colloid occurs primarily by changes in system conditions that affect the properties of the solvent.

This process is also known as phase separation, which is recognized by the appearance of turbidity, droplet formation or separation of liquid layers. It can be simple or complex, and may also occur as a result of the salting-out effect (Figure 6.16). The formation of the coacervate can occur either in a homogeneous medium, resulting in monolithic particles, or in a heterogeneous medium, wherein the coacervate is deposited on the surface of the dispersed phase, yielding reservoir-type systems.

In a simple coacervation process, the addition of a nonsolvent to the system causes the formation of a phase rich in polymers. It is based on the development of chemical reactions or the exchange of ions in the interface of an aqueous phase/organic phase, for hardening of the walls of the microcapsule. For example, some typical aqueous polymer dispersions and their nonsolvents may be gelatin and ethanol or acetone; agar

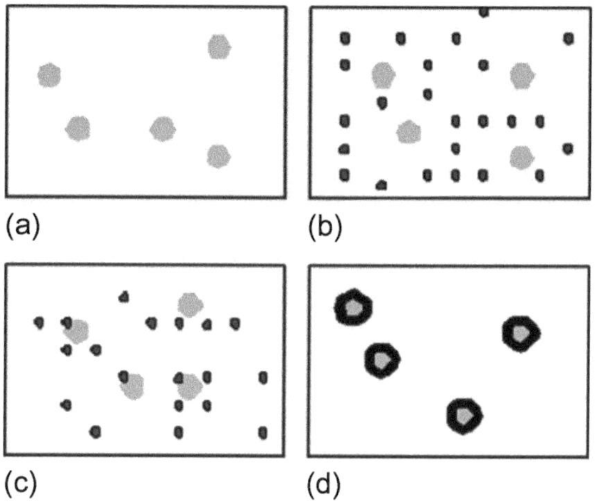

Figure 6.16 Microencapsulation by coacervation or phase separation: (a) dispersion of the active agent in the polymer solution; (b) beginning of coacervation; (c) gradual adsorption of micro-coacervates on the surface of the active agent; (d) coalescing of the micro-coacervate on the wall of the agent, curing of the coating and complete desolvation.

and acetone; pectin and dioxane or 2-propanol; methylcellulose and acetone; polyvinyl alcohol and 1-propanol; or fibrinogen and 1-propanol.

Complex coacervation is the result of mutual neutralization of two or more oppositely charged colloids in an aqueous solution. This method consists of dispersing the active component to be encapsulated in an aqueous solution of a polyelectrolyte, and depositing the micro-coacervate around the material to be encapsulated by adding an aqueous solution of a second oppositely charged electrolyte. The industrial importance of encapsulation by the complex coacervation technique is bigger due to the fact that this technique does not use chemical cross-linking agents.

Coacervation is a reversible process and, generally, the formation of microcapsules is a process of equilibrium. If the balance is destroyed, the phase separation is also destroyed.

Emulsification and solvent evaporation is a much-utilized method, consisting in solidifying the internal phase of an emulsion, yielding microparticles (Figure 6.17). The active agent may be dissolved or suspended in the polymer solution (coating material) using a volatile organic solvent. This phase is then emulsified in an immiscible dispersant aqueous medium containing a surfactant (emulsifier) that avoids agglomeration of the microparticles. The solvent is evaporated at low temperature and reduced pressure. The solubility of the polymer decreases as the solvent evaporates, resulting in microparticles that are separated by centrifugation or filtration. The method may yield microparticles from 5 to 5000 μm (Benita, 2006).

The structure of produced microparticles is essentially a matrix, where the active agent is dispersed in the polymer matrix, thus forming microspheres.

The fusion and emulsification method involves the melting of the polymer (coating material), followed by dissolution of the drug core or co-fusion of the two components.

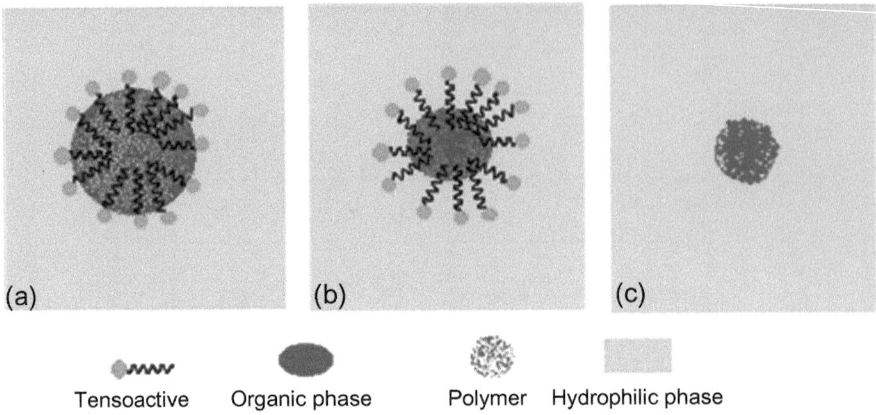

Figure 6.17 Schematic representation of the obtention of microspheres by emulsification and solvent evaporation.

Preparing an emulsion and the inner layer is solidified by decreasing the temperature of the system. The separation of the microparticles is carried out by filtration or centrifugation. The active agent needs to be thermoresistant, which is unlikely for most organic compounds.

Polymerization or the interfacial copolymerization method is based on properties of polymers. These are macromolecules that often have film-forming capacity and can form coatings, which are obtained by grouping reactions of the monomeric units. Thus, methods have been developed that promote the *in situ* formation of microcapsule walls using these polymers. Identical monomers undergo polymerization and different monomers undergo polycondensation.

Usually, this is a spontaneous chemical reaction and the active agent can be further adsorbed on the surface of the microparticles. Many times, nanoparticles are obtained with a great ability to target specific sites of therapeutic action. This reaction can occur in the external medium in which the active agent is dispersed as a solid or liquid, and the formed polymers diffuse to the interface, where they attach. It may also occur *in situ* at the interface, either spontaneously or by contact of the monomer diffused in one side and the catalyst in the other. It can still occur *in situ* at the interface by chemical condensation of two monomers, chemically dissimilar in opposite phases, reacting at the interface (Figure 6.18).

The main methods of microencapsulation by interfacial polymerization typically involve a chemical reaction between a diacyl chloride and an amine or an alcohol. The resulting polymer film can be polyester, polyurea, polyurethane or polycarbonate. Generally, reservoir-type devices are formed and the particle size is directly dependent on the diameter of the internal phase and the control of the reaction (initialization, propagation, and termination of the polymer chain).

Moreover, the interfacial polycondensation method is characterized by a chemical reaction between two chemically different monomers, which takes place in an organic phase-aqueous phase interface, giving rise to a polymer film. Briefly, an emulsion whose internal phase contains the active ingredient and one of the monomers is

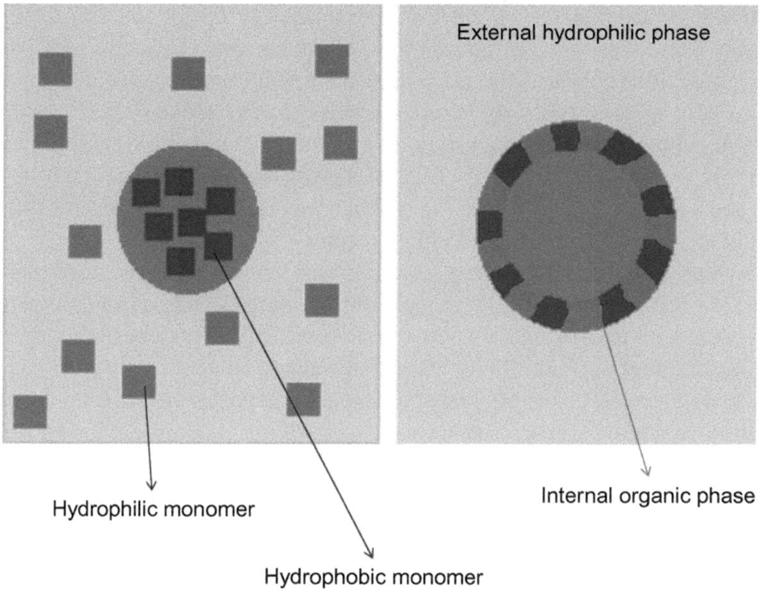

Figure 6.18 Schematic representation of microparticles obtained by interfacial polymerization.

prepared. The external phase contains a surfactant and other monomers. The monomers migrate to the interface where they react to form a polymer that encapsulates the inner phase. The initiator of the reaction may be a chemical or physical agent. Alternatively, the therapeutic agent may be added by interaction with a suspension of preformed microspheres. However, the first method allows the incorporation in high yield, while the second allows a nonspecific adsorption with low income.

By the gelling method, the core drug is dispersed in a sodium alginate dispersion, which is gelled by the addition of calcium chloride. The "droplets" are transferred to a polycation solution (e.g., polylysine) that displaces the calcium ion, hardening the membrane. The gel inside the microcapsule can be liquefied by the addition of sodium citrate, which shifts the remaining calcium ions. The method is particularly used to microencapsulate insulin and cellular material (Benita, 2006).

The atomization and drying method involves the drug dispersing in the molten coating material and the mixture atomizing in an environment with conditions to promote rapid solidification of the material. This process is accomplished by rapid evaporation of the solvent of the coating material or by freezing. The fast drying is achieved by atomizing the mixture in a stream of heated air or lyophilizing the frozen material.

Spray drying is a process widely used to dry solids by atomization (spray) from a dispersion of the drug and wall-forming material as fine droplets in a hot-air environment. The water then evaporates and the dried solid is obtained. Several points in the process are important, including the nucleus (amount of the wall material, viscosity, and temperature).

An aqueous solution, an organic solution or a mixture of the material to be coated can be used. Moreover, the design of the spray dryer may be important (structure of the drying chamber, inflow and air outlet, the drying temperature and the kind of collector). There are two types of nebulizer nozzles: the turbine model (nebulization is achieved by compressed air in a rotary disc or rotary atomization) and the atomizer model (fogging is done by compressed air at a fixed nozzle). Various factors such as exchange and heat losses, equity and the geometry of the nebulizer make it difficult to transform laboratory equipment for industrial scale.

The first step in encapsulation via spray drying is to disperse the core material in a concentrated solution (40–60% by weight) of the particle-forming material so as to obtain droplets with diameters between 1 and 3 μm. The polymers typically increase the viscosity at higher concentrations. Water is used as a solvent for this technique. The toxicity and flammability severely restrict the use of conventional organic solvents for encapsulation by spray drying. Furthermore, several groups are exploring this technique with organic media to produce microparticles with biodegradable polymers.

The dispersion sent to the chamber of the spray dryer is rapidly dehydrated, producing dry particles. These are captured by the collector (Figure 6.19). Microparticles produced in this manner typically have a diameter between 1.0 and 300 μm. They tend to have a spherical geometry and may be aggregated.

This microencapsulation technique has many advantages: it is a well-established technique; it is of comparatively low cost and is still in full development; it uses readily available equipment; it can be produced in large quantities; and many of the materials used by this technique are approved for use in foods and are soluble in water, favoring the dissolution of the microparticles and releasing the encapsulated material.

On the other hand, the spray-drying technique has problems and limitations. If water is the solvent chosen, the technique is limited to materials that are soluble or dispersible polymers in water. The encapsulation rate is typically around 20–30%,

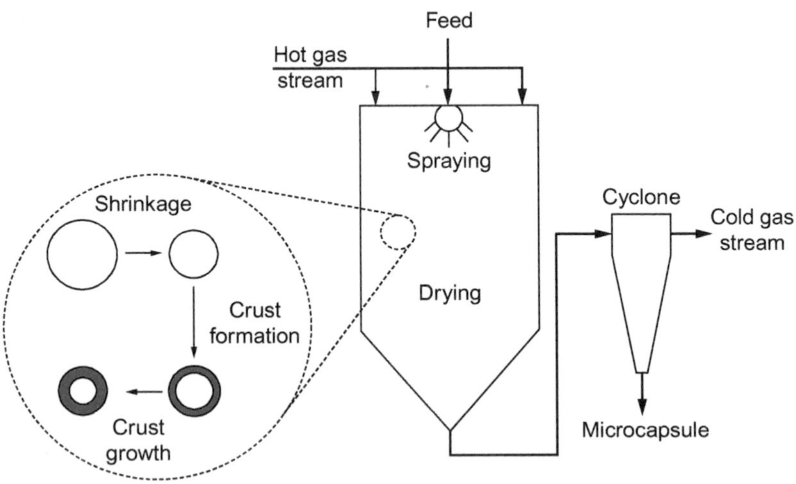

Figure 6.19 Schematic of a spray dryer and the process of microencapsulation.
Reprinted from Jamekhorshid, Sadrameli, and Farid (2014), with permission from Elsevier.

although there are protocols that report a rate between 50% and 60%. The unencapsulated material is another problem. The low boiling point of some polymers has been shown to be a persistent problem for the encapsulation, as they may volatilize in the chamber. Thus, the polymeric materials that can be used are polysaccharides (starch and gum arabic) or proteins (gelatin, albumin, casein).

Spray freezing is a feasible method for producing microparticles, proposed by Rogers, Hu, Yu, Johnston, and Williams (2002). A solution containing the drug and excipients is atomized under the surface of a liquid cryogen (e.g., liquid nitrogen). The droplets formed in the atomization solidify instantaneously (Figure 6.20). Frozen microparticles are then collected and lyophilized.

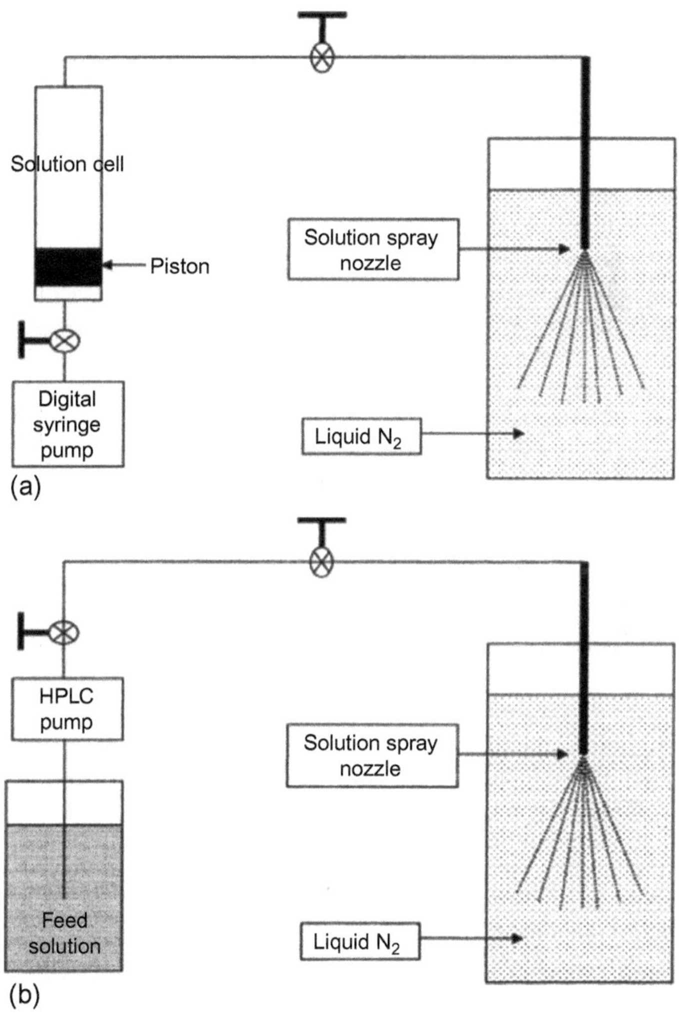

Figure 6.20 Schematic representations of spray freezing into liquid, utilizing liquid nitrogen as the cryogenic medium: (a) laboratory-scale process; (b) pilot-scale process.
Reprinted from Rogers et al. (2002), with permission from Elsevier.

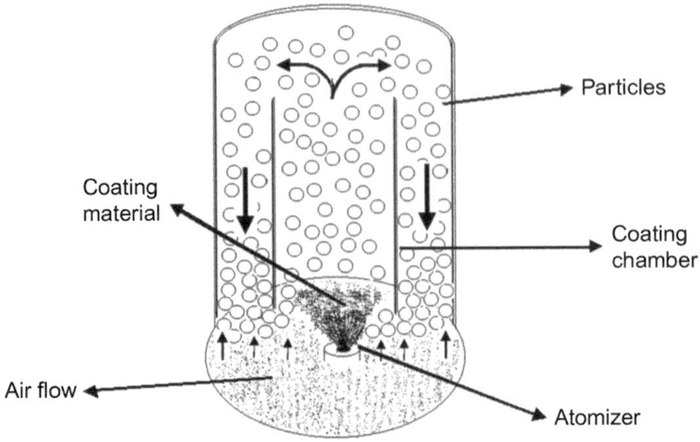

Figure 6.21 Schematic diagram illustrating the fluidized bed (Wuster).

The air suspending or fluid bed technique is based on dispersion of the active agent in the form of small particles (drug nucleus) and held suspended in an air stream through a fluidized bed, while the coater material is atomized in the particles in motion (Figure 6.21).

The drying is performed with the cyclized air in the coating chamber. It is applicable for solid or liquid cores adsorbed on solid supports. The size of the obtained microcapsules is in the range of 35–5000 μm, with the upper value having no size limit with respect to the technological process, but with respect to the microparticle system.

Various mechanical means are used for microencapsulation using centrifugal force. The multi-orifice spin uses centrifugal force to release the core by means of a coater membrane material, providing the mechanical effect of microencapsulation.

Another technique is centrifugal extrusion, by which the core material and the wall-forming material of the microcapsule (immiscible) are pumped through a spinning nozzle nebulizer. This produces a continuous column of the two fluids, spontaneously breaking up into spherical droplets. Each droplet contains a continuous central region bound by a liquid membrane. The way in which these droplets are converted into capsules is determined by the nature of the wall-forming material. If it has low viscosity when melted, which crystallizes quickly with cooling (e.g., wax or wax polymer), the droplets are converted into solid particles to leave the nozzle. The drugs most appropriate in this case are polar, because these are immiscible with most of the wall-forming materials, such as waxes.

Alternatively, the droplets emerging from the nozzle may have a wall formed by the hydrophilic polymer, which has the capacity to gel quickly. In this case, the droplets fall into a gelling bath, where they are gelled. A specific example of this type of microencapsulation is the particle being obtained by gelling an aqueous solution of sodium alginate in an aqueous solution of a calcium chloride bath. The non-polar active agents are suitable to be microencapsulated by this technique.

Rotational suspension is a technique of microencapsulation in which the material to be microencapsulated is dispersed in a polymeric solution. This dispersion passes through a rotating disc, which can be flat, tapered or filet. The individual particles are expelled by the rotating disc using centrifugal force, and the microcapsules are formed and solidified by cooling.

BRACE® is a set of patented processes proposed by Brandau (2002) for production of microspheres and microcapsules. The technique is based on the utilization of a vibrating nebulizer, producing uniform particle size. It is possible to obtain microspheres and microcapsules with a diameter from 30 to 8000 μm. For the preparation of microspheres, the drug can be dispersed, dissolved or emulsified in the matrix-forming material. For microcapsules, the core material may be an aqueous solution, emulsion, dispersion or melted material. The only restriction is that the material to be encapsulated produces no chemical reaction with the wall-forming material.

A wide variety of materials can be used for this process, provided they are in liquid form and display viscosity less than 10 Pas. The main materials are alginates, gelatin, agar, waxes, thermoplastics, metal oxides, polyethylene glycol, polyvinyl alcohol, polyacrylate, polystyrene, and methacrylate.

According to Brandau (2002), the process is based on pumping of the material to be microencapsulated through a nebulizer nozzle. A vibrating device induces the breakdown of uniform flow, forming drops (Figure 6.22). The droplets are solidified by a gelling system, by cooling, by chemical reaction or by drying during the falling in the chamber.

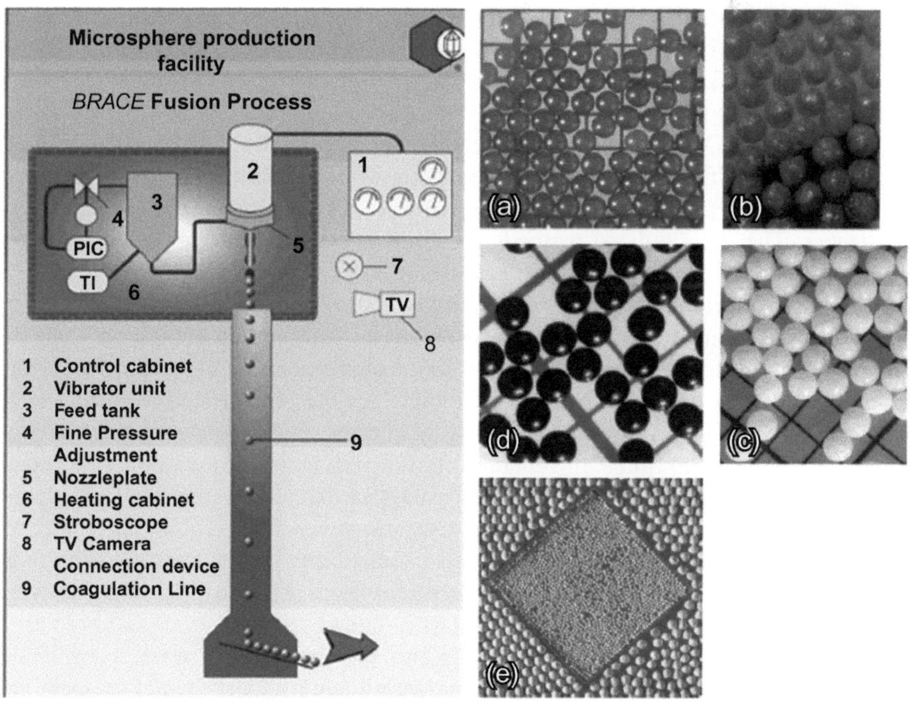

Figure 6.22 BRACE® microencapsulation process and different structures obtained: (a) cosmetic waxes; (b) agar-agar with oils; (c) polymer beads for combinatorial synthesis; (d) pharmaceutics encapsulated in wax; (e) inorganic microparticles as catalyst carriers
Reprinted from Brandau (2002), with permission from Elsevier.

Pan coating is a process widely used in microencapsulation, but its application is limited to solid active agents. It is considered essential that the core particles be at least 600 μm for the coating to be effective. This process is among the oldest industrial methods for obtention of small, coated particles or tablets, mainly in the preparation of prolonged-action dosage forms. The drug particles are tumbled in a pan or other device, while the coating material is generally applied slowly with the moving of the pan. Therefore, the temperature is decreased so that the coating material encloses the particles of the active agent (core) and then is solidified by cooling. Another procedure is to gradually apply the coating material to core particles tumbling in a vessel rather than being wholly mixed with the core particles from the start of encapsulation. The solvent used to disperse the coater material needs to be easily removed by hot air stream.

Drug encapsulation is possible using supercritical fluid, a state in which the substance has an intermediary behavior between the liquid and gas states. With characteristic conditions of temperature and pressure, the highly compressed gas shows several advantageous properties of both liquids and gases.

The most widely used are supercritical CO_2, alkanes (C_2–C_4) and nitrous oxide (N_2O). They have low hydrocarbon-like solubility for most solutes and are miscible with common gases such as hydrogen (H_2) and nitrogen.

The use of supercritical fluid in many industrial applications is dependent on their capacity to modify the density with a small change in temperature or pressure. Specifically, supercritical CO_2 is widely used for its low critical temperature (31 °C) and pressure (73.836 bar) values. Moreover, it is nontoxic, non-flammable, readily available, highly pure, and cost-effective.

Supercritical CO_2 has applications in encapsulation of active agents. A wide variety of materials that either dissolve (paraffin wax, acrylates, polyethylene glycol) or do not dissolve (proteins, polysaccharides) in supercritical CO_2 are used for encapsulating core substances. The most utilized methods are rapid expansion of supercritical solutions (RESS), gas anti-solvents (GAS), and particles from gas-saturated solutions (PGSS).

RESS is characterized to be a process by which the supercritical fluid containing the therapeutic agent and the encapsulating material is submitted at high pressure and then released at atmospheric pressure through a small nozzle. As the pressure drops suddenly, it causes desolvation and deposition of the material around the drug (core), forming a coating layer. Both the active agent and the coating material must be very soluble in supercritical fluids. In general, very few polymers with low cohesive energy densities (e.g., polydimethylsiloxanes, polymethacrylates, and polyethylene glycol) are soluble in supercritical fluids such as CO_2. However, co-solvents can be utilized to increase the solubility of the polymers. In some cases nonsolvents are used (e.g., ethanol), increasing the solubility in supercritical fluids, but the polymeric materials do not dissolve at atmospheric pressure.

GAS is also called the supercritical fluid anti-solvent (SAS) process, in which the supercritical fluid is added to a solution of coating material and the active agent and maintained at high pressure. This leads to a volume expansion of the solution that causes super saturation, precipitating the solute. The solute should not dissolve in the mixture of solvent and supercritical fluid, but must be soluble in the liquid solvent. In addition, the liquid solvent must be miscible with the supercritical fluid.

This process is unsuitable for the encapsulation of water-soluble ingredients, as water has low solubility in supercritical fluids. It is also possible to produce submicron particles using this method.

The PGSS process is reasoned on the mixture of the active agent (core) and coating materials in supercritical fluid at high pressure. During this process, supercritical fluid penetrates the encapsulating material, causing swelling. When the mixture is heated above the glass transition temperature, the polymer liquefies. However, as the pressure decreases, the encapsulating material deposits onto the active agent. It is clear that both the drug and encapsulating materials may not be soluble in the supercritical fluid.

Another strategy using supercritical fluid technology is the utilization of preformed microparticles for the entrapment of active agents. When the pressure is decreased, the microparticles shrink and return to their original shape and entrap the ingredients (Jain, 1997).

6.6 Micro/nanoemulsioned systems

6.6.1 Microemulsions

Microscopic emulsion-like structures in a mixture of water, cetyl trimethyl ammonium bromide (cationic surfactant), oil phase and alcohol (hexanol) was first proposed by Hoar and Schulman in 1943 (Hoar & Schulman, 1943). In 1959, these structures were finally confirmed by solidifying the oil phase through staining, and visualized by electron microscopy (Schulman, Stoeckenius, & Prince, 1959).

Coarse emulsions are thermodynamically unstable. They display average drop size, growing continuously with time so that phase separation ultimately occurs under gravitational force, and their formation requires the input of work. Unlike for these emulsions, the concept of microemulsion is based on a mixture of oil, water and surfactants, whereby spontaneous formation occurs, rendering a transparent liquid dispersion. Danielsson and Lindamn (1981) defined microemulsion as "a system of water, oil, and amphiphile which is a single optically isotropic and thermodynamically stable liquid solution."

6.6.1.1 Types

According to the properties of the components, microemulsions can form over a wide range of oil–water–surfactant compositions. The chemical and physicochemical properties of each component are fundamental for obtention of microemulsioned systems. In addition, microemulsions are normally obtained in narrow, specific concentration ranges. To evaluate and find these ranges, pseudo-ternary phase diagrams (ratios between water, oil and surfactant-co-surfactant) are produced (Figure 6.23). The region of microemulsion is characterized by compositions that produce a very low interfacial tension at the water and oil interface, the formation of a highly fluid interfacial surfactant film, and the penetration and association of the molecules of the oil phase with the interfacial surfactant film.

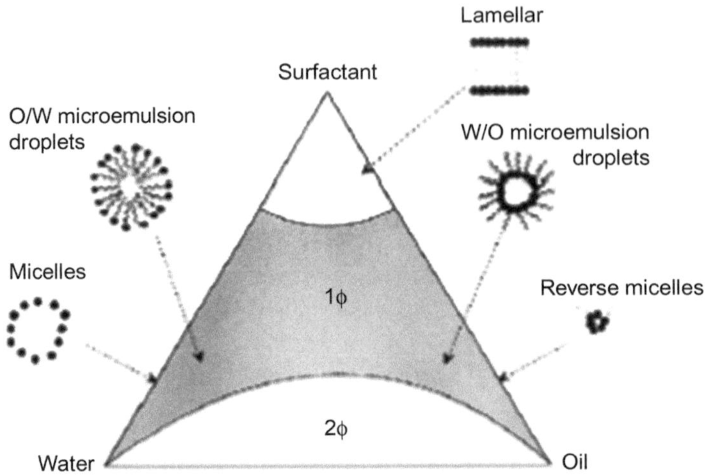

Figure 6.23 A hypothetical pseudo-ternary phase diagram of an oil/surfactant/water system emphasizing the microemulsion phase.
Reprinted from Lawrence and Rees (2012), with permission from Elsevier.

When water, oil, and surfactants are mixed, a wide variety of structures and phases can be formed (emulsions, anisotropic crystalline cubic or hexagonal phases, lamellar structures, and microemulsions) and are dependent on the ratio of the components. These different structures may be recognized by visual analysis of their physical appearance. Emulsions are non-transparent, cubic phases and lamellar structures are highly viscous, and crystalline phases are revealed by the use of polarized light during inspection.

Therefore, they are differentiated from actual microemulsions, which are translucent and dynamic systems displaying the interface continuously and spontaneously fluctuating (Figure 6.24). The difference in size between a swollen micelle (approx. 5–140 nm) and a fine emulsion (approx. 100–600 nm) is not fully defined, and a droplet of the microemulsion may disappear within a fraction of a second, whilst another droplet may form spontaneously in the system.

Microemulsions show behavior different from emulsions and, in many situations, the emulsion terminologies "oil-in-water (o/w)" and "water-in-oil (w/o)" are not applicable to these systems. When the amounts of oil and water in the system are comparable, there is an interpenetration of domains of equilibrium bicontinuous structures, producing more complicated structures. Both water and oil domains extend over macroscopic distances, and the surfactant exerts the important function to form an interface of rapidly fluctuating curvature, but in which the net curvature is near zero.

Winsor (1948) classified the microemulsioned systems into four types:

- Type I (the surfactant is preferentially soluble in water and o/w microemulsion form)
- Type II (the surfactant is mainly in the oil phase and w/o microemulsion form)
- Type III (a three-phase system where a surfactant-rich middle-phase coexists with both excess water and oil surfactant-poor phases)
- Type IV (a single-phase (isotropic) micellar solution that forms upon addition of a sufficient quantity of amphiphiles)

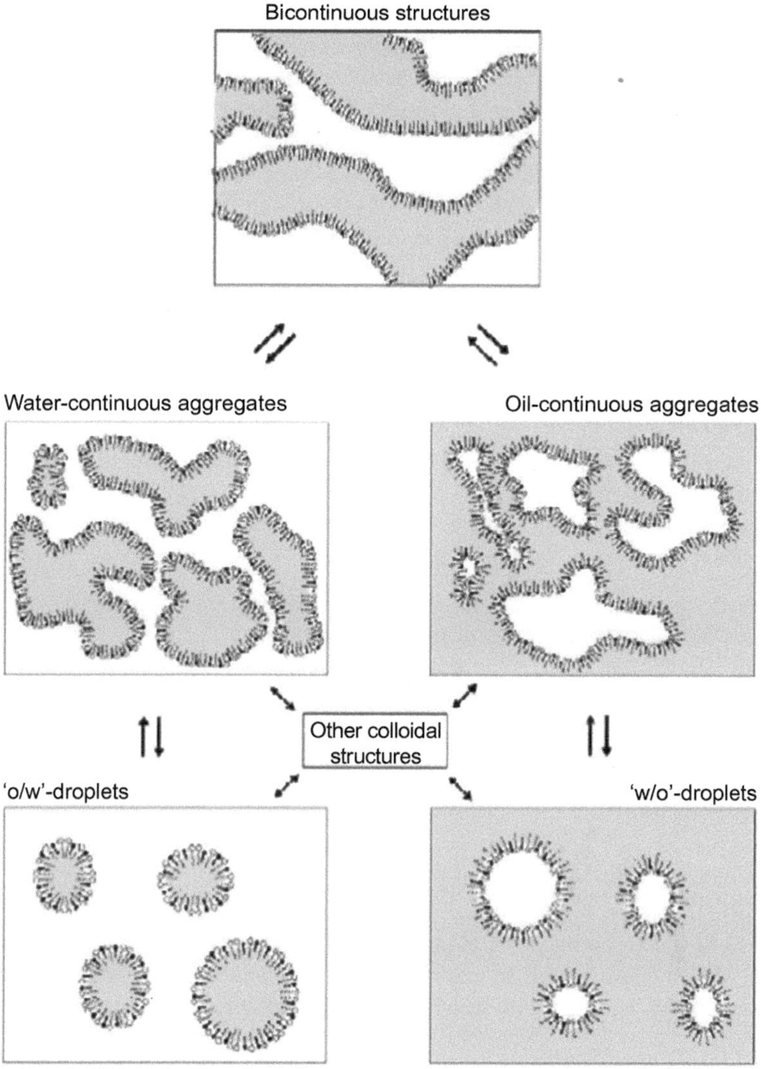

Figure 6.24 Structures of basic dynamic microemulsion obtained by mixture of oil (gray), water (white) and surfactant/co-surfactant interfacial film, and plausible transitions between the structures by increase of oil fraction (clockwise from left to right) and water fraction (anticlockwise from right to left), respectively.
Reprinted from Kreilgaard (2002), with permission from Elsevier.

Therefore, microemulsion comprises a number of different structures, most of which have little in common with emulsion. Depending on surfactant type and sample environment, types I, II, III, or IV form preferentially, the dominant type being related to the molecular arrangement at the interface. These transitions and structural formations are not fully understood in terms of mechanism. However, it is determined that flexibility of the surfactant film is an important factor. In this sense, a more flexible surfactant film will enable the existence of different structures (e.g., aggregates and bicontinuous structures) and a wide variety of compositions of microemulsions (Kreilgaard, 2002).

The introduction of a short-chain co-surfactant to the surfactant film enables the decreasing of the interfacial tension and fluidization of the interfacial surfactant film. On the other hand, a single surfactant system is possible because it has lipophilic chains sufficiently short and/or fluidizing groups like those containing unsaturated bonds. In the case of microemulsion, the attainment of very low interfacial tension is an essential requirement and the use of a second amphiphile (surfactant), usually a medium-chain length alcohol called a co-surfactant, is generally necessary.

Microemulsions are transparent, having low viscosity, thermodynamic stability, and the ability to form spontaneously, without the necessity to provide mechanical energy (e.g., high shear equipment and heating), and their microstructures are independent of the order of addition of the excipients. Moreover, they provide ultralow interfacial tensions and large interfacial areas, as well as the ability to concentrate and localize significant amounts of both oil- and water-soluble active agents within the same isotropic medium.

Three theories have been used to explain microemulsion formation and stability (Prince, 1967; Kreilgaard, 2002):

(1) Interfacial or mixed film theory, in which the negative interfacial tension due to high film pressure is responsible for the formation of microemulsions
(2) Solubilization theory
(3) Thermodynamic treatments

In general, the extent to which the surfactant lowers the surface tension of the oil–water interface and the change in the entropy of the system has importance on the free energy of microemulsion formation:

$$\Delta G_{\mathrm{f}} = \gamma \cdot \Delta A - T \cdot \Delta S \tag{6.7}$$

where ΔG_{f} is the free energy of formation, γ is the surface tension of the oil–water interface, ΔA is the change in interfacial area on microemulsification, ΔS is the change in entropy of the system that is effectively the dispersion entropy, and T is the temperature.

When the microemulsion system is formed, the change in ΔA is very large due to the large number of very small droplets formed, and the entropy is raised. However, there are also expected to be favorable entropic contributions arising from other dynamic processes, such as surfactant diffusion in the interfacial layer and monomer-micelle surfactant exchange. The surfactants have as their main function to reduce the interfacial tension sufficiently, lowering the energy. Thus, a negative free energy of formation is achieved when large reductions in surface tension are accompanied by significant favorable entropic change. In these situations, the microemulsification process is spontaneous and the resulting dispersion is thermodynamically stable (Lawrence & Rees, 2012).

These systems have attracted much interest in drug delivery for oral, parenteral, pulmonary, ocular, and topical delivery. They provide more rapid and efficient oral absorption of active agents than solid dosage forms (enhancement of bioavailability of active agents), the increasing of cutaneous drug delivery of both hydrophilic and lipophilic agents compared to conventional vehicles, and the unique potential application in the development of artificial red blood cells and targeting of cytotoxic active agents to cancer cells. During the last two decades, a lot of research work has been carried out on microemulsion systems for providing novel solutions to overcome the problems of poor aqueous solubility of highly lipophilic drug compounds and provide reproducible bioavailability.

The design and development of new microemulsions with the intention of enhancing the efficacy of existing drugs is an ongoing process in pharmaceutical research. In this sense, another type of system studied is the self-microemulsifying drug delivery system (SMEDDS), which is not a microemulsion, but is related. A SMEDDS is composed of a mixture of oil, surfactants and the therapeutic agent, which when introduced into the body is rapidly dispersed to form droplets of the same size of microemulsions and to behave *in vivo* much the same way as o/w microemulsions.

On the other hand, microemulsions require much larger amounts of surfactants for their preparation, which restricts the choice of acceptable components. Regardless of this, the use of environmentally responsive drug delivery systems is a very important strategy. Specifically for microemulsions, it is possible to develop systems displaying phase changes after administration triggered by changes in pH, ionic strength, or temperature.

A solution of lecithin in isopropyl myristate enables the phase transformation of a reverse micellar solution to a lamellar liquid crystal. The reverse micellar solution in contact with a biological aqueous phase changes phase, and the controlled release of the anti-inflammatory fenoprofen occurs (Muller-Goymann & Hamann, 1993).

Thermosetting microemulsions as delivery systems for periodontal anesthesia (lidocaine and prilocaine) have been developed using block copolymer liquid microemulsion (Scherlund, Malmsten, Holmqvist, & Brodin, 2000). When the system is administered into the periodontal pocket, the microemulsion viscosity increases, as a consequence of the temperature increasing.

In addition, the sol-gel transitions are useful to be administered by parenteral route. When the formulation is injected, it undergoes subcutaneous gelation. In this sense, thermoresponsive polymeric block copolymer micelles based on poly(*N*-isopropylacrylamide) and poly(butylmethacrylate) and containing adriamycin were also developed (Chung et al., 1999).

The development of cubosomes was possible using mixtures of glyceryl monoleate, water and poloxamer 407 (a triblock PEP–PPO–PEO copolymer). The obtained microemulsion is a cubic bicontinuous phase, which could be dispersed in water, yielding the cubosomes (Gustafsson, Ljusberg-Wahren, Almgren, & Larsson, 1996).

From an industrial point of view, it can be easily scaled up, considering the relative cost of commercial production. Microemulsion can also be used for cosmetic purposes and drug targeting. Nowadays, researchers' work is focused on the production of safe, efficient and more compatible microemulsion constituents that will further enhance the utility of this novel delivery system. Therefore, microemulsions have proven it possible to formulate preparations suitable for most routes of administration (oral, topical, parenteral, ophthalmic, brain and displayed potential and nasal) as multipurpose drug delivery systems. Their advantages include thermodynamic stability, optical clarity and ease of penetration.

6.6.2 Nanoemulsions

A nanoemulsion can be considered a conventional emulsion that contains very small particles. It is defined as a dispersed system containing submicrometer-size droplets, displaying stability against sedimentation or creaming and a transparent or translucent optical aspect (similar to that of microemulsions). However, this nanosized dispersion is a non-equilibrium system that may undergo flocculation, coalescence, and/or Ostwald ripening. Despite this, it is possible to select an appropriate composition and preparation method in order to achieve nanoemulsion with high kinetic stability.

Sometimes there is a misconception in the scientific literature between nanoemulsions and microemulsions. The International Union of Pure and Applied Chemistry (IUPAC) states, "In an emulsion, liquid droplets and/or liquid crystals are dispersed in a liquid" (International Union, 1972). When the word "dispersed" is interpreted as non-equilibrium and opposite of "solubilized," the IUPAC definition is not applied to microemulsions and micellar systems.

In this sense, a nanoemulsion is a mini-emulsion, submicron-size emulsion or ultrafine emulsion that may possess a relatively high kinetic stability, even for several years. Therefore, nanoemulsions can be classified as dispersed systems that possess droplet sizes varying from 20 nm up to 100–500 nm and represent the extreme lower limit of mini-emulsions. In contrast to nanoemulsions, most mini-emulsions are comprised of droplets in the size range from 100 nm to 1 μm.

In addition, nanoemulsions may be of the o/w or w/o type, depending on whether the oil is dispersed as droplets in water, or vice versa. Their optical properties range from opaque to nearly transparent, which drives increasing research activity in nanoemulsion.

It is possible to disperse on phase in another immiscible phase without using a surfactant. However, the formed system is highly unstable to droplet coalescence and a surfactant (or surfactants) is needed to facilitate the formation of the nanoemulsion and to ensure its kinetic stability during storage.

In this sense, an o/w nanoemulsion is structured by the non-polar tails of the surfactant molecules protruded into the hydrophobic core formed by the oil phase, while the polar head groups of the surfactant molecules protrude into the surrounding aqueous phase (Figure 6.25). The structure of the w/o nanoemulsion is the opposite.

6.6.2.1 Methods of preparation

To prepare nanoemulsions, it is commonly necessary to use high-energy methods of dispersion. Mechanical devices (e.g., high-shear stirrers, high-pressure homogenizers, and ultrasound generators) are able to produce intense disruptive forces. There is a direct relationship between the applied energy and the size of droplets: the higher

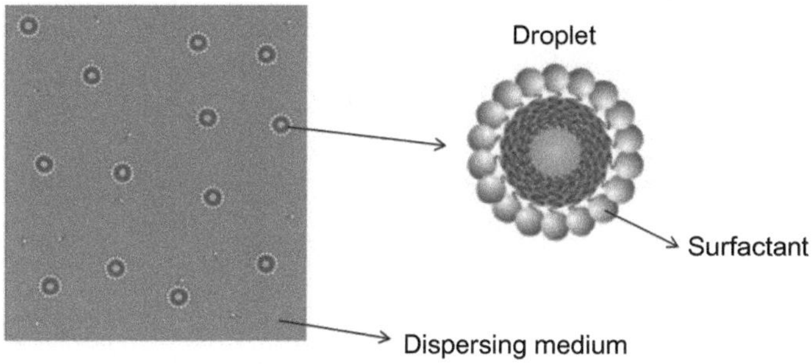

Figure 6.25 Schematic representation of structure of a nanoemulsion. The droplet size is from 20 nm up to 100–500 nm.

the energy input, the smaller the droplet size. This necessity of applying high energy constitutes a problem and leads this method of preparation to be seen as cost-inefficient, especially considering that only a small amount (around 0.1%) of the energy produced is used for emulsification (Tadros, Izquierdo, Esquena, & Solans, 2004).

On the other hand, there are low-energy methods used to prepare nanoemulsioned systems. In this case, only simple stirring is needed, the internal chemical energy of the system is used, and the method is considered more energy efficient.

It is also very important to consider the possibility of using both methods of preparation to achieve nanoemulsions with similar droplet sizes, depending on the system and composition variables.

Emulsion making was one of first applications of the ultrasound. Nowadays, there is a variety of types of ultrasonic devices. Cavitation is the main phenomenon of ultrasonically induced effects, which is the formation and collapse of vapor cavities in a flowing liquid. Two mechanisms are proposed for ultrasonic emulsification. First, the application of an acoustic field produces interfacial waves, resulting in the dispersion of the oil phase in the continuous phase in the form of droplets. The second one is that the application of ultrasound causes acoustic cavitation, causing the formation, and subsequent collapse of microbubbles by the pressure fluctuations of a simple sound wave, which creates extreme levels of highly localized turbulence. Therefore, the turbulent micro-implosions break up primary droplets into submicron size. Because the emitted sound field is typically inhomogeneous in most ultrasonic devices, it is necessary to recirculate the emulsions through the region of high power so that all droplets experience the highest shear rate.

The development of high-pressure homogenizers in the past 15 years makes it possible to explore nanoemulsioned systems. By this methodology, a coarse emulsion is prepared using a high-speed mixer and is then fed into the input of a high-pressure valve homogenizer. As it enters the gap between the valve and the valve seat (Figure 6.26), the flow velocity is increased rapidly. Homogenization is completed in the area between the valve and the seat, where the emulsion experiences a combination of intense disruptive forces that cause the larger droplets to be broken down to smaller ones. The combination of two theories, turbulence and cavitation, explain the droplet size reduction during the homogenization process.

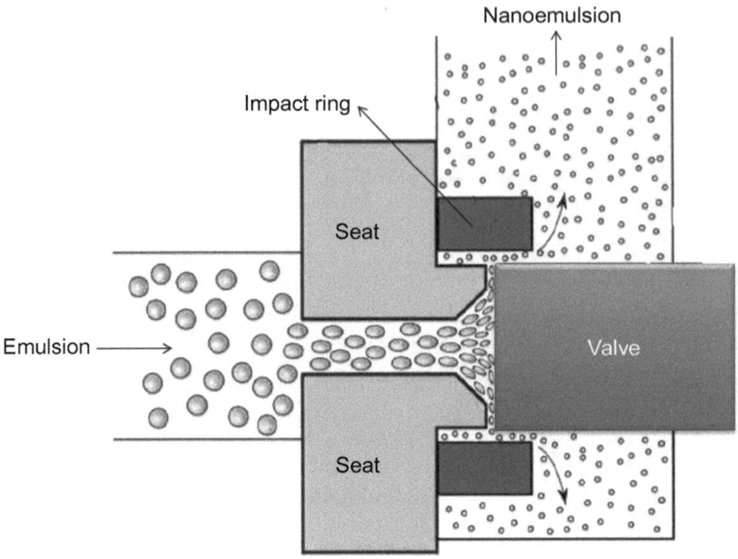

Figure 6.26 Schematic representation of production of microemulsion using a high-pressure homogenizer.

The high velocity gives the liquid high energy in the homogenizer valve and generates intense turbulent eddies. The droplets are torn apart by these eddy currents, resulting in a reduction in droplet size. Considerable pressure dropping across the valve occurs together with cavitation, generating further eddies and droplet disrupture. The degree of cavitation can be regulated by decreasing the gap size, which increases the pressure drop. The obtained nanoemulsions generally show diameters as small as 100 nm, using a suitable and sufficient amount of surfactant.

Nowadays, microfluidization has been proposed to prepare nanoemulsions. It is a patented mixing technology that makes use of a device called a microfluidizer, which uses a high-pressure positive displacement pump (500–20,000 psi). The components of formulation are forced through the interaction chamber, consisting of small channels (microchannels). The product flows through the microchannels to an impingement area, resulting in very fine particles of submicron range. First, the two phases are mixed in an inline homogenizer to yield a coarse emulsion which is introduced into a microfluidizer, rendering the nanoemulsion. The coarse emulsion is passed through the interaction chamber of the microfluidizer repeatedly, until the desired particle size is obtained. The bulk emulsion is then filtered through a filter under nitrogen to remove large droplets, resulting in a uniform nanoemulsion.

Independent of preparation method, nanoemulsion formation is a process of self-emulsification, which is triggered by the rapid diffusion of surfactant and/or solvent molecules from the dispersed phase to the continuous phase without involving a change in the spontaneous curvature of the surfactant.

The pharmaceutical industry uses the spontaneous emulsification method to obtain o/w nanoemulsions as carriers for lipophilic active agents in a hydrophilic media, with low energy costs. They are usually referred to in the literature as self-nanoemulsifying drug delivery systems (SNEDDS) (Solans & Solé, 2012).

Phase inversion methods are also utilized. They are characterized by spontaneous changes in the surfactant curvature during the emulsification process from negative to positive (to obtain o/w emulsions) or vice versa (for w/o emulsions). These phase inversion methods are classified as phase inversion temperature (PIT) and phase inversion composition (PIC) methods if emulsification is triggered by a change in temperature or composition, respectively.

To make a "stable" nanoemulsion reproducibly, a large number of factors must be controlled, like an appropriate composition, controlling the order of addition of the components, and applying the shear (when it will be the case) in a manner that effectively ruptures droplets. The dispersed phase molecules must be essentially insoluble in the continuous phase so that Ostwald ripening does not occur rapidly, despite the very high Laplace pressures. The choice of surfactant and other components needs to be appropriated to avoid the formation of lyotropic liquid crystalline phases. Systems containing short-chain alkanes, alcohols, water and surfactants are known to form these phases. The third requirement is that the continuous phase have a significant excess of surfactant to enable a new surface area of the nanoscale droplets to be rapidly coated during emulsification, thereby inhibiting shear-induced coalescence. This excess is generally in the form of surfactant micelles in the continuous phase. These micelles dissociate into monomers that rapidly adsorb onto the surfaces of newly created droplets. Finally, considering the high-energy method of fabrication, the high shear must be applied to rupture microscale droplets into nanodroplets. The stress level must reach the Laplace pressure of droplets with the desired size, usually in the range of 10–100 atm (Mason, Wilking, Meleson, Chang, & Graves, 2006).

6.6.2.2 Applications

Nanoemulsions have attracted enormous interest due to their wide range of applications, namely in the pharmaceutical, food, cosmetic, and chemical industries.

The ability of nanosized systems to dissolve large amounts of hydrophobics, along with their mutual compatibility and ability to protect the therapeutic agents from hydrolysis and enzymatic degradation, make them ideal vehicles for the purpose of parenteral transport. Moreover, the frequency and dosage of injections can be reduced throughout the drug therapy period, as these emulsions guarantee the release of drugs in a sustained and controlled mode over long periods of time. Despite the nanoemulsion instability, the lack of flocculation, sedimentation, and creaming, combined with a large surface area and free energy, offers advantages over emulsions of larger particle size, for parenteral route of administration. Their very large interfacial area positively influences drug transport and delivery, along with targeting them to specific sites (Lovelyn & Attama, 2011).

Nanoemulsion systems have the advantage of a high capacity to solubilize drug compounds and to retain the drug in solution even after dilution. For example, nanoemulsions can improve the oral bioavailability of poorly water-soluble drugs.

The self-nanoemulsified drug delivery systems can also be nanosized. They can be isotropic mixtures of oil, surfactant, co-surfactant and drug that form fine o/w nanoemulsions (approx. 100 nm) when introduced into aqueous phases under gentle agitation (e.g., gastrointestinal (GI) tract) (McClements, 2012). These systems improve the oral bioavailability of poorly water-soluble drugs by different mechanisms, including improved drug solubilization and protection against physicochemical and enzymatic degradation. In addition, the small size of these droplets increases the interphase between the lipophilic droplet and the aqueous medium of the gut, facilitating a homogeneous and wide distribution of the drug along the GI tract.

For example, self-nanoemulsified drug delivery systems using Cremophor® EL and a mixture of ethanol and glyceryl dioleate, mixed with hydroxypropylmethylcellulose, were developed by Gao et al. (2003) for the oral delivery of paclitaxel. This formulation displayed a fourfold higher bioavailability of the anticancer drug than when formulated as Taxol®.

Therefore, nanoemulsification constitutes a very important strategy to modify the delivery of therapeutic agents, enabling the obtainment of delivery systems for the safe and effective delivery of poor oral bioavailability agents. Nanoemulsions have great potential for application to human therapy by increasing the efficiency of oral administration.

6.7 Liposomes

Liposomes are close spherical bilayer vesicles composed of phospholipids. Despite the fact that there is a range of vesicle-based preparations and lipid complexes available, not all of them can be defined as liposomes. In consequence of phospholipid organization, the liposome presents two distinct regions: one hydrophilic or aqueous core surrounded by one or more concentrically arranged bilayer membranes (Figure 6.27). The aqueous core space ranges from about 0.03 to 10 μm in diameter.

The bilayer membrane can be composed of natural or synthetic lipid molecules. However, usually phospholipids are used to prepare liposomes. They can entrap hydrophilic and lipophilic active agents within their aqueous core and the lipid bilayers, respectively. Moreover, liposomes can be prepared in a range of size with single or multiple bilayers.

Alec Bangham and colleagues published the first description of swollen phospholipid systems (Bangham, Standish, & Watkins, 1965). From their study, the basis for model membrane systems was established and a variety of enclosed phospholipid bilayer structures consisting of single bilayers, initially termed "bangosomes" and afterward "liposomes," were described. However, Gregory Gregoriadis defined the liposome as a vesicle system that could entrap active agents and be used as a drug delivery system. Later, other researchers showed that liposomes could change the *in vivo* distribution of entrapped active agents.

In consequence of their structure, liposomes were the first of the parenteral particulate drug delivery systems to be approved for clinical use. Nowadays, there are several liposome formulations used in clinical practice.

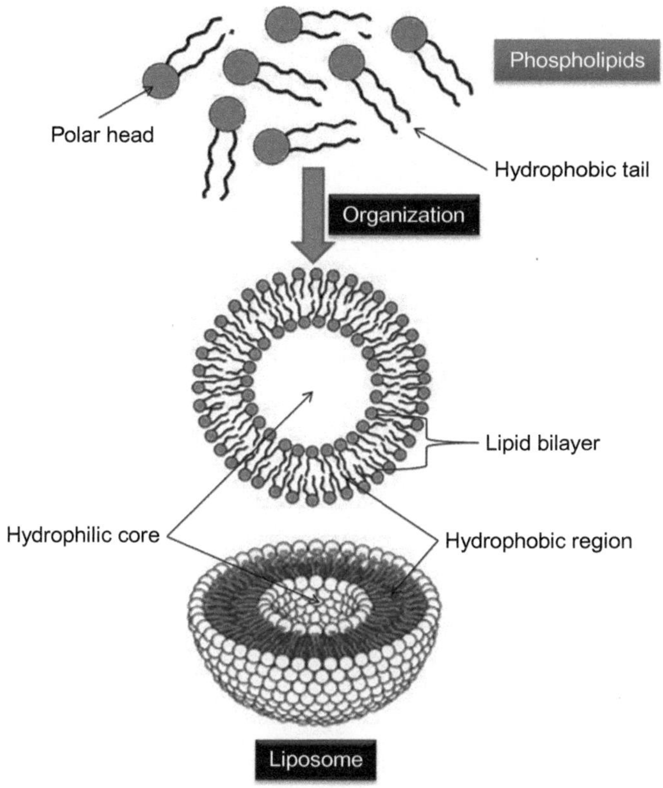

Figure 6.27 Schematic representation of liposome structure showing the phospholipids, their organization, the lipid bilayer and the three-dimensional structure.

They have the ability to protect the entrapped active agent from the external environment and change the biodistribution of the active by providing passive targeting of the active or active targeting through the addition of targeting groups to their structure surface. Moreover, through modification of the bilayer composition, the delivery of the entrapped active agent can also be controlled.

The first medicine containing liposomes was Pevaril-Lipogel®, an antifungal containing econazole, which hit the market in 1988. Nowadays, a wide range of active agents has been studied in the liposomal form, and targeting is the main strategy for these studies.

6.7.1 Classification

Liposomes are classified according to their size and number of bilayers, which are very important to control the delivery of the active agent. Moreover, numerous configurations are possible for liposomes, including spheres and cylinders.

Spherical liposomes can be unilamellar or multilamellar and are often designated LUV (large unilamellar vesicle), SUV (small unilamellar vesicle) and MLV (multilamellar vesicle) (Table 6.5). The MLVs may have an "onionskin" structure of several layers.

Table 6.5 **Morphological and size classification of liposomes**

Classification	Morphology	Diameter (nm)	Structure
MLV (multilamellar vesicles)	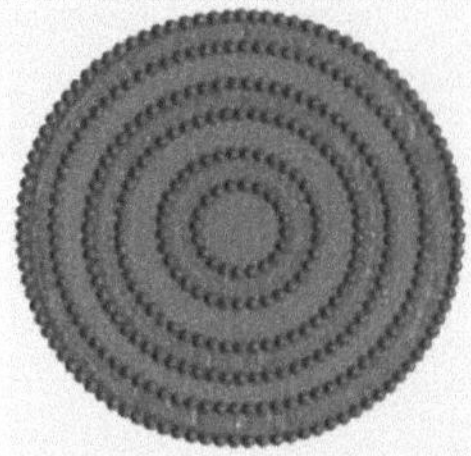	100–5000	Varied number of bilayers
SUV (small unilamellar vesicles)	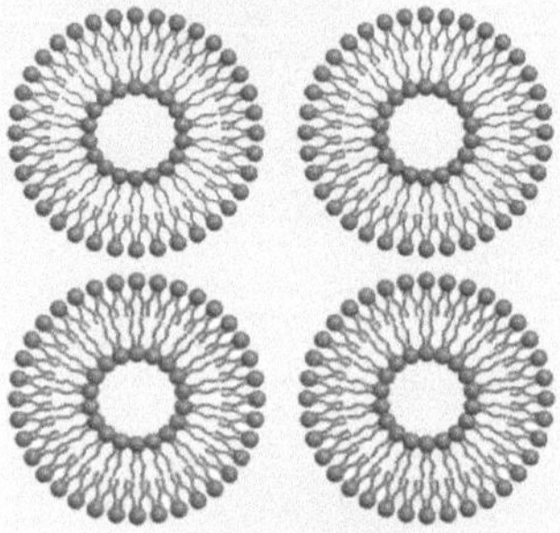	<100	Unilamellar small vesicles filled with aqueous solution
LUV (large unilamellar vesicles)	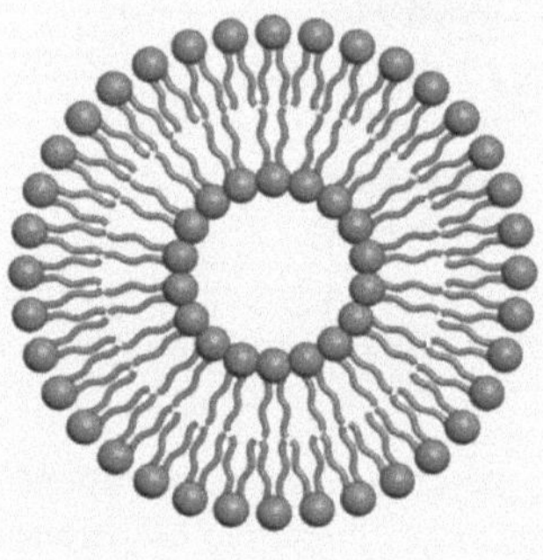	100–2000	Unilamellar large vesicles filled with aqueous solution

In terms of size, the smaller liposomes generally range from 0.02 to 0.2 μm, while the large ones from about 0.2 μm to more than 10 μm.

6.7.2 *Chemical composition and preparation*

Liposomes can be composed of different phospholipids, but the main ones are neutral phospholipids (phosphatidylcholine). Additional constituents are cholesterol (40–50%) to increase the stability in biological fluids, and charged substances (5–20%) to prevent aggregation (e.g., acidic lipids like phosphatidylserine, phosphatidylglycerol, phosphatidylinositol, phosphatidic acid, and cardiolipins).

The phospholipids of liposomes are amphipathic (they possess both a polar head and a non-polar tail) and their behavior is similar to the hydrophil–lipophil balance and wedge-orientation theories of emulsification.

The non-polar tail is composed of fatty acids containing generally 10–24 carbon atoms, and the aqueous head may contain phosphoric acid bound to a water-soluble portion. In this sense, phosphatidylcholine (lecithin) is a lipid that has been used extensively in the production of liposomes.

When these phospholipids are exposed to a polar solvent (e.g., water) and line up, the fatty acid tails associate together as the lipophilic phase and the aqueous head groups associate toward the bulk water phase. Therefore, according to the system and its solubility, the active may be trapped in the aqueous compartments (if water-soluble) or in the lipophilic bilayers (if oil-soluble).

The preparation of liposomes involves the dispersion of phospholipids and other components in a solvent, which is further removed. The solvent evaporation leaves a thin film of the lipid on the walls of the container. The addition of an aqueous solution of the active agent to the container conduces to the displacement of the lipid film from the container walls, forming spheres or cylinders, trapping the aqueous solution inside. Afterward, the liposomes can be collected, sized and used.

This methodology is termed "film hydration" and produces MLV with low granulometric quality. One variation of this method uses vigorous emulsification, by ultrasound, of an organic phase and an aqueous phase. After emulsification, the organic solvent is evaporated, generally by nitrogen insufflations, giving MLV liposomes.

All morphological types of liposomes (e.g., MLV and LUV) can be submitted to ultrasound after their preparation in order to produce SUV liposomes. Another strategy is to use the extrusion method by which the MLVs are forced through small orifices under high pressure.

The freeze drying of SUVs can result in MLVs. The aqueous phase containing the active agent is a mixture with a pre-obtained SUV suspension. The system is submitted to liophylization, leading to the rupture of preformed SUVs. The rehydration of liophylized originates MLVs containing the encapsulated active agent.

The injection of non-polar solvents produces SUVs or oligolamellar vesicles and uses as solvent ether or fluorocarbon. The lipids are dissolved in the organic solvent and the dispersion is injected in an aqueous solution of the active agent to be encapsulated. The vesicles are formed by contact with the aqueous medium and by evaporation of the organic solvent. One variation of this method is the injection of water-soluble solvents in which solutions of lipids in water-miscible solvents like ethanol, glycerol and polyglycols are utilized. At the contact with the aqueous phase, the solvent mixture and the lipid solubility decrease, forming the vesicles.

Vesicles greater than 100 nm (LUVs) show advantages like high encapsulation efficiency of hydrosoluble drugs, lipid economy and reproducible profiles of drug delivery. However, this type of liposome is the most difficult to prepare. Detergent remotion is an alternative method by which aqueous dispersions containing phospholipids and detergents are prepared. The last one is slowly removed by dialysis. The micelles with a lot of phospholipids coalesce and LUVs are obtained. Moreover, it is possible to use the method of evaporation in reverse phase. In this case, an emulsion (w/o) composed of buffer (containing the active agent) and phospholipids in excess of the organic phase are prepared. The organic phase is removed under reduced pressure, producing the LUVs.

There are many variations and improvements of the described methods, which enable the production of liposomes with quality and quantity suitable for industries. Size control of liposomes and encapsulation rate are two fundamental characteristics that need to be controlled during production.

6.7.3 Applications

Liposomes have been studied to modify and control the delivery of active agents to be administered for a wide range of routes (e.g., parenteral, topical and by inhalation). Liposome-encapsulated actives are protected and delivered intact to various sites and can be released when the liposome is destroyed. They have the ability to be selectively absorbed by tissues rich in reticuloendothelial cells, such as the liver, spleen, and bone marrow (Allen & Cullis, 2013). This characteristic can serve as a targeting to other tissues, and cells of the body are protected from the drug, decreasing the drug's toxicity.

Lipophilic and/or hydrophilic drugs can be carried by liposomes without chemical modification of the carrier. However, the charge, size, and other characteristics can be altered depending on the active agent and the intended use of the product.

Nowadays, liposomes constitute advanced vesicle carriers for drug delivery with enhanced size, composition, and stability. There is a range of liposomal products in the market.

Doxorubicin, a cytotoxic anthracycline antibiotic isolated from *Streptomyces peucetius* var. *caesius* agent and used to treat advanced ovarian cancer, breast cancer, and AIDS-related Kaposi's sarcoma, is available in sterically stabilized liposomes (80–100 nm) composed of pegylated distearoyl phosphatidyl ethanolamine, hydrogenated soy phosphatidylcholine, and cholesterol. Doxil®, a liposome injection containing doxorubicin hydrochloride, consists of the active agent encapsulated in stealth liposomes to be administered intravenously.

Stealth liposomes are pegylated and, therefore, protected from detection by the mononuclear phagocyte (Figure 6.28). They have increased the blood circulation time, displaying a half-life of approximately 55 h in humans.

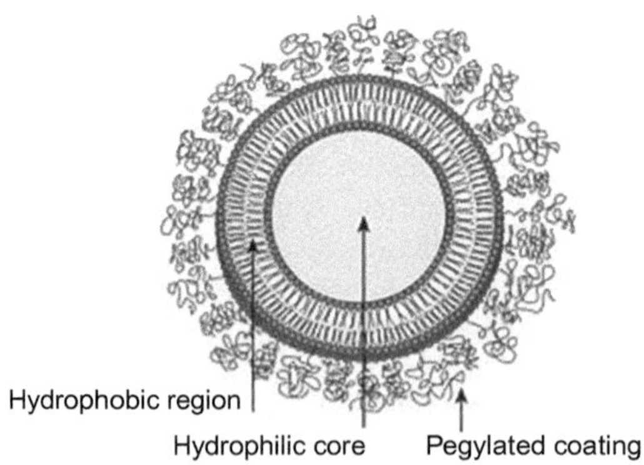

Figure 6.28 Schematic representation of a pegylated liposome (Stealth® liposome).

Liposomes containing doxorubicin with size of 150–190 nm and composed of egg phosphatidylcholine and cholesterol are also available as the first-line treatment of metastic breast cancer in women. In addition, liposomes (45 nm) containing daunorubicin and composed of distearoyl phosphatidylcholine and cholesterol are available to treat Kaposi's sarcoma.

Cytabarine, an active agent for the intrathecal treatment of lymphomatous meningitis, is available in the liposomal form. This formulation is characterized to be composed of multivesicles (size from 3 to 30 μm) of dioleoylphosphatidylcholine, cholesterol, triolein, and dipalmitoylphosphatidylglycerol.

Another very important liposomal formulation in the market is the liposome containing amphotericin B for the treatment of fungal infections and leishmaniasis. The vesicles have size less than 100 nm and are composed of soy phosphatidylcholine, cholesterol and distearoylphosphatidylglycerol.

Liposomes have also been used as an important strategy to deliver vaccines. Vesicles containing formalin inactivated hepatitis A virus are available and are named virosomes, with size of 150 nm, composed of influenza virus surface antigens phosphatidylcholine and phosphatidylethanolamine. Purified influenza hemagglutinin glycoprotein and neutaminidase is available encapsulated in liposomes (virosomes), similar to that previously described, with the indication to the influenza vaccine.

Liposomes have also been used as a strategy to modify the delivery of active agents used in photodynamic therapy. Verteporin, a photosensitizer, is available in liposomes, composed of dimyristoyl phosphatidylcholine and egg phosphatidylglycerol, for the photodynamic therapy of macular degeneration.

6.8 Niosomes

During the last 20 years, the use of nanotechnology has rapidly increased. This has enabled the development of nanocarriers each more specialized. The ability to manipulate molecules and supramolecular structures for beneficial functions led to the development of structures suitable for biopharmaceutical and therapeutic applications.

As previously discussed for liposomes, phospholipids are biodegradable and nontoxic amphiphiles displaying important properties to produce vesicle carriers. However, they show low physical and chemical stability in aqueous suspensions and need to be stabilized. Therefore, non-ionic surfactants have been studied to obtain vesicular carriers of active agents.

They are known as niosomes and, in terms of structure and physical properties, are similar to liposomes. In general, niosomes are prepared following the same procedures and under the same variety of conditions as those described for liposomes.

The vesicle structures may be unilamellar or multilamellar, and they are able to overcome some disadvantages associated with liposomes. Different from phospholipids, surfactants are easily derivatized, they possess a higher versatility to form vesicular structures, and they have lower costs than phospholipids.

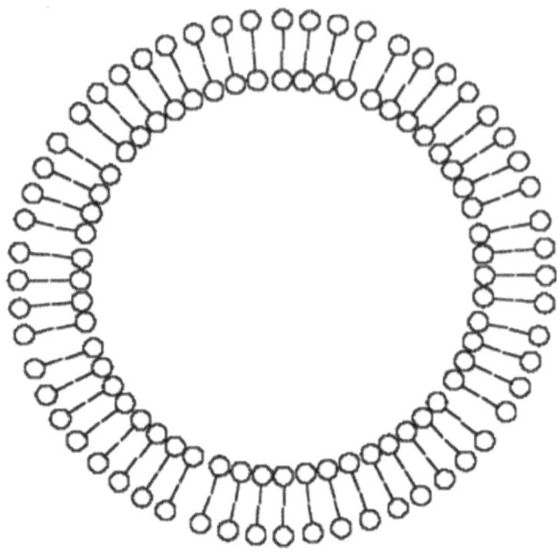

Figure 6.29 Schematic representation of a niosome: (O) hydrophilic head group; (–) hydrophobic tail.
Reprinted from International Uchegbu and Suresh (1998), with permission from Elsevier.

These non-ionic surfactant-based vesicles are formed from the self-assembly of non-ionic amphiphiles in aqueous media, resulting in closed bilayer structures. They are capable of encapsulating both hydrophilic and lipophilic substances. The hydrophilic substance can be entrapped within the vesicular aqueous core or adsorbed on the bilayer surfaces, while the lipophilic substances are encapsulated by their partitioning into the lipophilic domain of the bilayers (Figure 6.29).

6.8.1 Classification

According to their size, niosomes can be divided into SUV (sizing from 10 to 100 nm), LUV (from 100 to 300 nm) and MLV containing more than one bilayer (Figure 6.30). Most niosomes are classified as colloidal structures (nano or submicron size range).

In addition, there are different types of specialized niosomes:

- Proniosomes (They are liquid crystalline compact niosome hybrids that, upon hydration, form niosomes. They are a strategy to reduce the physical stability problems of niosomes like aggregation, fusion and leaking and can improve the characteristics of dosing, distribution, transportation and storage showing.)
- Ethosomes (Surfactant ethosomes are composed of phospholipids, with a relatively high concentration of ethanol, water, and some surfactants. They have displayed enhanced skin permeation.)
- Elastic niosomes (They are composed of non-ionic surfactants, ethanol and water, showing superior capability to penetrate a compound through intact skin by passing through pores in the stratum corneum, due to the flexibility of their structure.)
- Polyhedral niosomes (They are niosomes composed of mixtures of determined anionic surfactants, cholesterol and ethers of cholesterol, producing vesicles with polyhedral shape.)
- Discomes (disc-shaped vesicles)
- Aspasomes (ascorbyl palmitate vesicles)

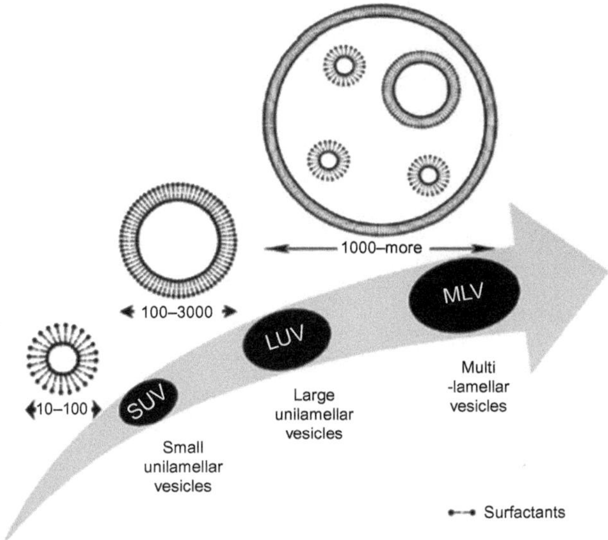

Figure 6.30 Schematic structure of niosomes, classifying according to morphology and size. Reprinted from Moghassemi and Hadjizadeh (2014), with permission from Elsevier.

6.8.2 Composition and preparation

Niosomes are generally composed of non-ionic surfactants, cholesterol and/or dicetyl phosphate (DCP).

According to their characteristics, the non-ionic surfactants can form structures in solution without charge in their polar heads. The non-ionic amphiphiles used in niosomes are classified in four categories: alkyl esters, alkyl amides, alkyl ethers, and esters of fatty acids. The most used are Span (20, 40, 60, 65, 80, and 85), Tween (20, 40, 60, 65, 80, and 85), and Brij (30, 35, 52, 56, and 58).

The selection of surfactant to prepare the niosome is dependent on the hydrophilic–lipophilic balance and critical packing parameter.

The addition of a charged molecule to the bilayer is one of the methods used to stabilize the niosome. Therefore, DCP and phosphatidic acid are known as negatively charged molecules, while stearylamine (SA) and cetylpyridinium chloride are positively charged molecules that are both commonly used for preventing aggregation of niosomes. DCP is a prevalent additive used as a charge inducer, imparting a negative charge on the surface of niosomes. This negative charge stabilizes their bilayers or achieves an electrophoretic mobility similar to that of erythrocytes, as in the case of encapsulated hemoglobins.

It is necessary to take care about the amount of the charged molecule to be added to niosomes. Normally, concentrations from 2.5 to 5 mol% are used. Higher concentrations can inhibit the formation of niosomes.

These types of vesicles are prepared by various methods, like hand shaking, thin-film hydration, proniosome technology, dehydration–rehydration, freeze and thaw, heating, microfluidization, sonication, reverse phase evaporation, ether injection, and bubble (without the use of organic solvent).

It is also possible to prepare proniosomes that are stable precursors of niosomes. They are dry formulations in which each particle is covered with a thin film of dry carrier (e.g., maltodextrin, sucrose stearate, or mannitol) with non-ionic surfactants. They show some important advantages, such as the reduction of niosome physical instability such as aggregation, fusion and leaking, and the increase of drug entrapment efficiency.

6.8.3 Applications

Niosome production was first started from the cosmetic industry, and then potential applications of niosomes in drug delivery were explored. They have been one of the illustrious vesicles in all vesicular systems, being the focus of great attention for potential drug delivery.

Important therapeutic agents have been encapsulated into niosomes (e.g., doxorubicin, monoxide, ovalbumin, insulin, oligonucleotide, hemagglutinin, DNA vaccine, α-interferon, bovine basic pancreatic inhibitor, and many others). In consequence, niosomes can be applied as anticancer, anti-inflammatory, anti-asthma, antimicrobial, anti-amyloid, antioxidant, anti-Alzheimer, and antibacterial.

According to their size and structure, these vesicles can be administered by routes like intramuscular, intravenous, ocular, subcutaneous, oral, pulmonary, intraperitoneal, and transdermal.

These characteristics of niosomes lead to numerous advantages. They are osmotically active and chemically stable, and they have long storage time compared to liposomes. It is easy to form and modify their surface because of the functional groups on their hydrophilic heads. The vesicles have high compatibility with biological systems and low toxicity, and are biodegradable and non-immunogenic.

Niosomes have the ability to encapsulate lipophilic active agents into vesicular bilayer membranes and hydrophilic active agents in aqueous compartments, improving the therapeutic performance of the agents by protecting them from the biological environment. This results in better availability and controlled drug delivery by restricting the drug effects to target cells in targeted carriers and delaying clearance from the circulation in sustained drug delivery. In addition, the access to raw materials is convenient, and the niosomes exhibit high patient compliance because of the water-based suspension.

Therefore, niosomes constitute drug delivery systems that have demonstrated to yield promising therapeutics.

6.9 Floating systems

Oral controlled drug delivery systems display considerable therapeutic advantages, such as ease of administration, patient compliance and flexibility of formulation. Important advances have been observed in these systems, but problems like the poor absorption window of active agents through the gastrointestinal tract have limited the success of the dosage forms.

The development of drug delivery systems to be administered to the gastrointestinal tract involves the difficulty to overcome the inability to confine the system to the desired area.

The gastrointestinal tract is a complex system, comprising different environments in which the active agent can suffer influence. Therefore, absorption of the active agent is a complex procedure and is subject to many variables.

Although there are various ways to prolong the release and action time of active agents that are orally administered, many physiological limitations (e.g., incomplete drug release, transit time and permanence in the gastrointestinal tract) are observed. The gastrointestinal residence time is dependent on many factors, such as size and density of dosage form, food intake, type of food, sleep, body posture, exercise, and others.

In this context, the strategy to prepare systems that can remain in the gastric region for several hours, and hence significantly prolong the gastric residence time of therapeutics, improves bioavailability, reduces drug waste, and improves solubility for active agents that are less soluble in a high pH environment.

The dosage form and the fed/fasted state of the stomach influence the gastric emptying of active agents. Normal gastric residence times usually range between 5 min and 2 h. In the fasted state, the electrical activity in the stomach, the interdigestive myoelectric cycle or migrating myoelectric complex governs the activity and, hence, the transit of dosage forms. It is characterized by a period of no contractions (40–60 min, phase I), a period of intermittent contractions (20–40 min, phase II), a period of regular contractions at the maximal frequency that travel distally (10–20 min, phase III or housekeeper wave) and a period of transition between phase III and phase I (0–5 min, phase IV).

The successful development of a gastric-retentive drug delivery system is dependent on some strategies. Low-density systems, hydrodynamically balanced systems (HBSs), raft systems and floating drug delivery systems are techniques currently used (Bardonnet, Faivre, Pugh, Piffaretti, & Falson, 2006).

Floating systems have the ability to prolong the gastric residence time, favoring the prolonged release of active agents that are absorbed in the stomach, or retarding the residence time in the gastrointestinal time and allowing the prolonged release into the upper intestinal portion. The system floats on the gastric contents, the active agent is released at the desired rate and, after drug release, the residual system is emptied from the gastric site.

6.9.1 Mechanism of floating

The gastric retention time of systems is dependent on the density and size of the dosage form. Moreover, there are other variables that are important (e.g., food intake, nature of the food, posture, age, sex, sleep, disease state of the individual, and administration of drugs such as prokinetic agents).

Dosage forms having a density lower than that of gastric fluid experience floating behavior and hence gastric retention. A density of < 1.0 g/ml is required to exhibit a floating property. However, the floating tendency of the dosage form usually decreases as a function of time, as the dosage form gets immersed into the fluid as a result of the development of hydrodynamic equilibrium (Timmermans & Moes, 1990).

The size of the system is an important factor affecting gastric retention. Floating systems remain buoyant on gastric fluids and are less likely to be expelled from the stomach, compared with the non-floating systems, which lie in the antrum region and are propelled by peristaltic waves (Timmermans & Andre, 1994).

The presence or absence of food in the stomach influences the gastric retention time of the system. The presence of food increases the retention time and increases the absorption of the active agent by allowing it to stay at the absorption site for a longer time.

In general, females showed comparatively shorter mean ambulatory gastric retention time than males. In addition, the gastric emptying in men is higher than in women.

Posture exerts no effect on gastric retention time. However, the floating and non-floating systems behave differently. In the upright position, the floating systems float to the top of the gastric contents and remain for a longer time, showing prolonged retention time. On the other hand, the non-floating dosage forms settle to the lower part of the stomach and undergo faster emptying with the peristaltic contractions. However, in the supine position, the floating units are emptied faster than non-floating units of similar size (Timmermans & Moes, 1994).

Floating systems are able to maintain their low apparent density, while the polymer hydrates and builds a gelled barrier at the outer surface. The active agent is released progressively from the swollen system, which remains floating on the gastric contents without affecting the intrinsic rate of emptying because its bulk density is lower than that of the gastric contents.

6.9.2 Classification

Based on the mechanism of buoyancy, floating systems have been developed using two distinct technologies: effervescent systems and non-effervescent systems.

The technology of the effervescent systems includes use of gas-generating agents, carbonates (e.g., sodium bicarbonate) and other organic acids (e.g., citric acid and tartaric acid) present in the formulation to produce carbon dioxide (CO_2) gas. Carbon dioxide generation reduces the density of the system, making it float on the gastric fluid. An alternative is the incorporation of a matrix containing a portion of liquid, which produces gas that evaporates at body temperature. Therefore, effervescent systems can be further classified into gas-generating and volatile liquid/vacuum systems.

Gas-generating systems are subdivided into:

(a) Intragastric single-layer floating tablets. They are composed of CO_2-generating agents and the active agent within the matrix tablet. The system receives water from the gastric fluid, swells, and displays bulk density lower than gastric fluids. In consequence, the system floats in the stomach, unflattering the gastric emptying rate for a prolonged period. The active agent is released at a desired rate for a prolonged period and is expelled from the stomach (Figure 6.31).

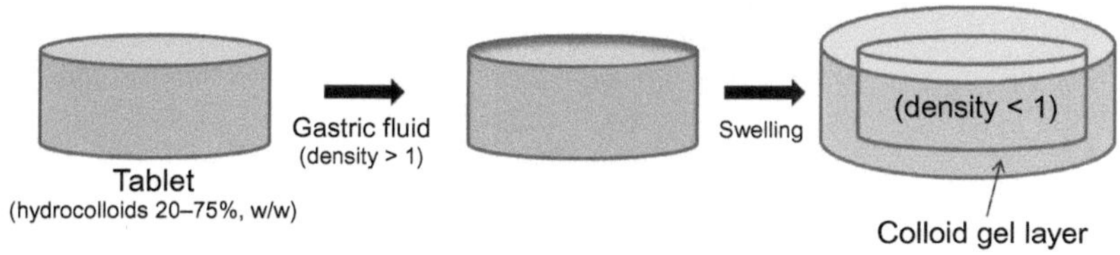

Figure 6.31 Schematic representation of the operating mechanism of an intragastric floating tablet.

(b) Intragastric bilayer floating tablets. They are compressed tablets containing an immediate release layer and a sustained release layer, and the operating mechanism is similar to the anterior (Figure 6.32).

(c) Multiple unit-type floating pills. These systems are sustained release pills surrounded by double layers. The inner layer consists of effervescent agents, while the outer layer is a swellable membrane. The water from the gastric fluid swells the pills like balloons, which float because they acquire low density. Carbon dioxide is generated within the system (Figure 6.33).

The volatile liquid/vacuum-containing systems are divided into:

(a) Intragastric floating gastrointestinal drug delivery systems. They have a floatation chamber, which may be a vacuum or filled with air or a harmless gas, while the reservoir of active agent is encapsulated inside a microporous compartment (Figure 6.34).

(b) Inflatable gastrointestinal drug delivery systems. These systems possess an inflatable chamber, which contains liquid ether that gasifies at body temperature to cause the chamber to inflate in the stomach. These systems are fabricated by loading the inflatable chamber with a drug reservoir, which can be a drug-impregnated polymeric matrix, then encapsulated in a gelatin capsule. After oral administration, the capsule dissolves to release the active agent reservoir together with the inflatable chamber. The inflatable chamber automatically inflates and retains the drug reservoir in the gastric fluid (Figure 6.35).

(c) Intragastric osmotically controlled drug delivery system. This type of device possesses an inflatable floating support based on osmotic pressure contained in a biodegradable capsule. In the stomach, the capsule quickly disintegrates to release the intragastric osmotically controlled drug delivery device. The inflatable support inside forms a deformable hollow polymeric bag that contains a liquid that gasifies at body temperature to inflate the bag. The osmotic pressure-controlled drug delivery device consists of two components: a drug reservoir compartment and an osmotically active compartment (Figure 6.36).

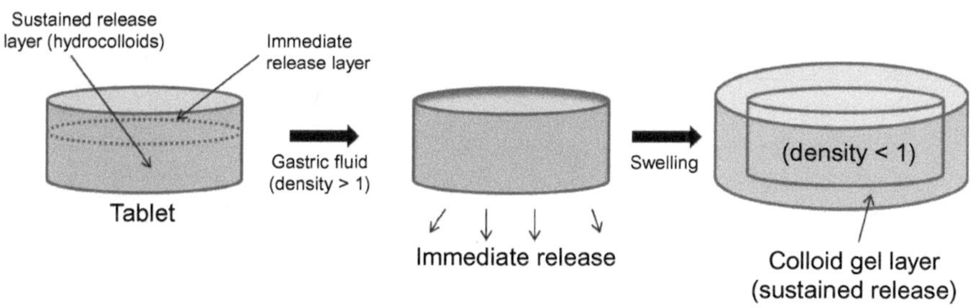

Figure 6.32 Schematic representation of the operating mechanism of an intragastric bilayer floating tablet.

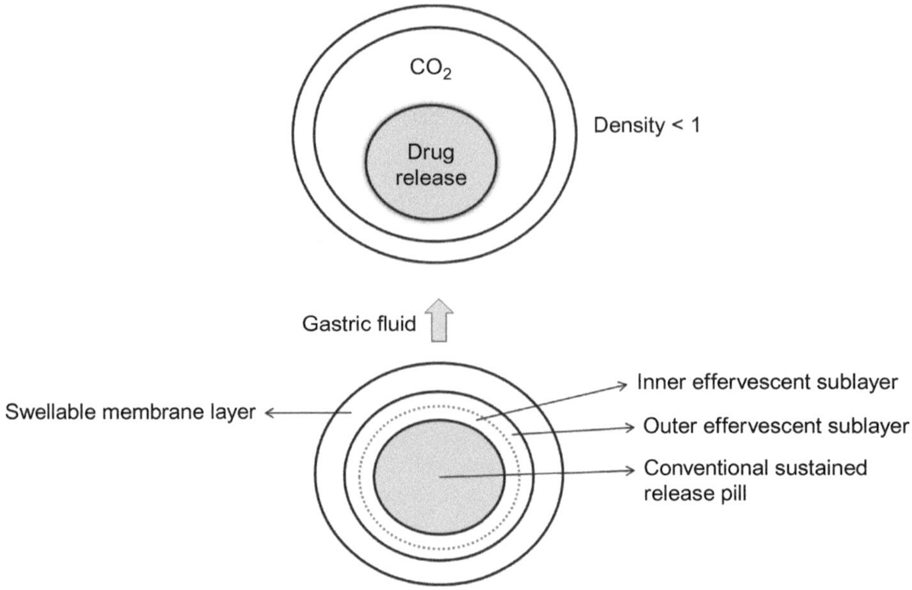

Figure 6.33 Schematic representation of the operating mechanism of multiple unit type floating pills.

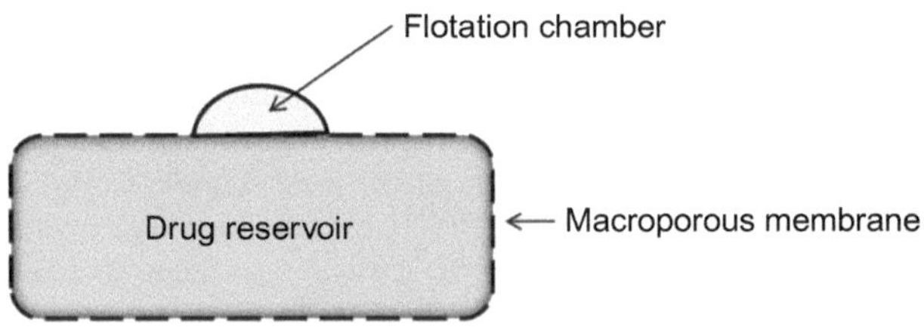

Figure 6.34 Schematic representation of the operating mechanism of an intragastric floating drug delivery device containing a flotation chamber.

Non-effervescent systems are normally composed of highly swellable cellulose-type hydrocolloids, polysaccharides and matrix-forming material (e.g., polycarbonate, polyacrylate, polymethacrylate and polystyrene, as well as bioadhesive polymer such as chitosan and Carbopol).

Swelling is an important strategy in this case. A colloidal gel barrier is a strategy to prepare non-effervescent floating systems with gel-forming hydrocolloids. This type of system possesses a high level of one or more gel-forming highly soluble cellulose-type hydrocolloids as hydroxypropyl cellulose, HEC, hydroxypropyl methyl cellulose, polysaccharides and matrix-forming polymers such as polycarbophil, polyacrylate and polystyrene. The water from the gastric fluid hydrates the hydrocolloids, and a colloidal gel barrier around its surface is formed.

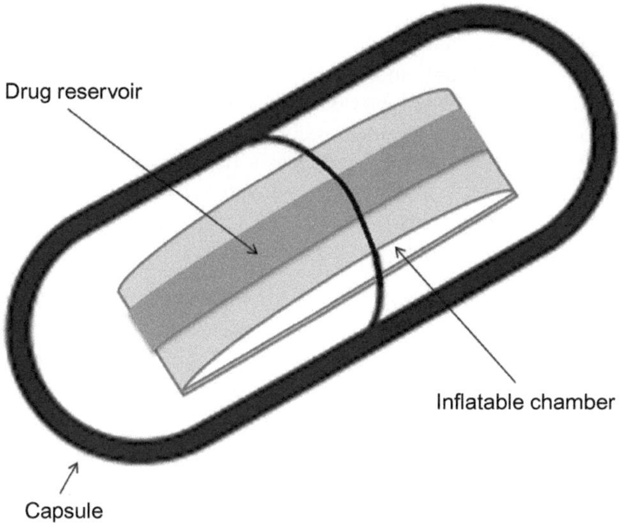

Figure 6.35 Schematic representation of the operating mechanism of an intragastric inflatable drug delivery device.

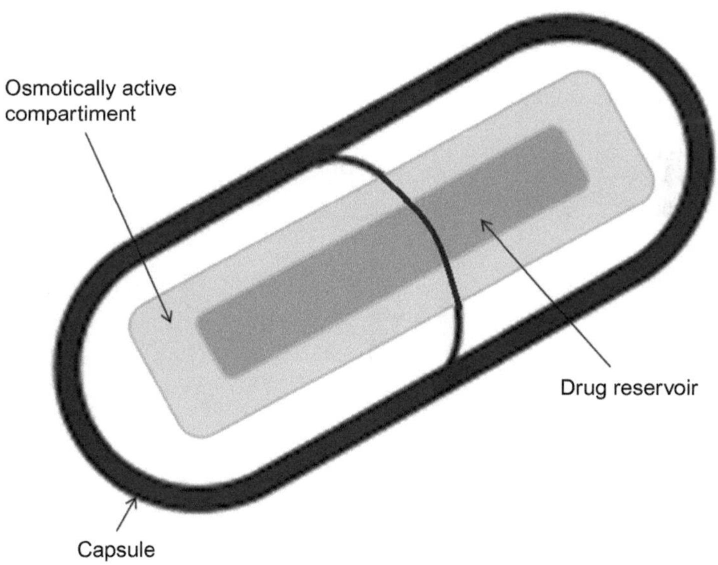

Figure 6.36 Schematic representation of the operating mechanism of an intragastric osmotic controlled drug delivery device.

(a) HBSs: These are single-unit dosage forms, containing one or more gel-forming hydrophilic polymers. Hydroxypropyl methyl cellulose is the most common used excipient, although HEC, hydroxypropyl cellulose, sodium carboxymethyl cellulose, agar, carrageenans, or alginic acid are also used. The polymer is mixed with the active agent and usually administered in a gelatin capsule. The capsule rapidly dissolves in the gastric fluid, and hydration and swelling of the surface polymers produces a floating mass. The release of the active agent is controlled by the formation of a hydrated boundary at the surface. Continuous erosion of the surface allows water penetration to the inner layers, maintaining surface hydration and buoyancy (Figure 6.37) (Bardonnet et al., 2006).

Another non-effervescent floating type is the microporous compartment system, in which a reservoir of active agent is encapsulated inside a microporous compartment with pores along its top and bottom walls. The reservoir is completely sealed to prevent any direct contact of the gastric surface with the undissolved active agent. Air is entrapped in a flotation chamber, allowing the system to float over the gastric content in the stomach.

Hollow microspheres and beads are other types of non-effervescent floating systems. Hollow microspheres are also known as microballoons, and are prepared by emulsification and the solvent diffusion method. Enteric acrylic polymer dispersed in the ethanol:dichloromethane solution of the active agent was poured into a stirred aqueous solution of polyvinyl alcohol at 40 °C. This process conduces to the formation of the gas phase in dispersed polymer droplets by evaporation of dichloromethane formed in an internal cavity of the polymeric microsphere containing the active agent. (Figure 6.38). The microspheres can float continuously over the surface of gastric media for more than 12 h *in vitro*.

6.9.3 Composition

Between the proposed floating systems, those based on the association of oils with swelling polymers, those that use porous materials, and the microballoons obtained by

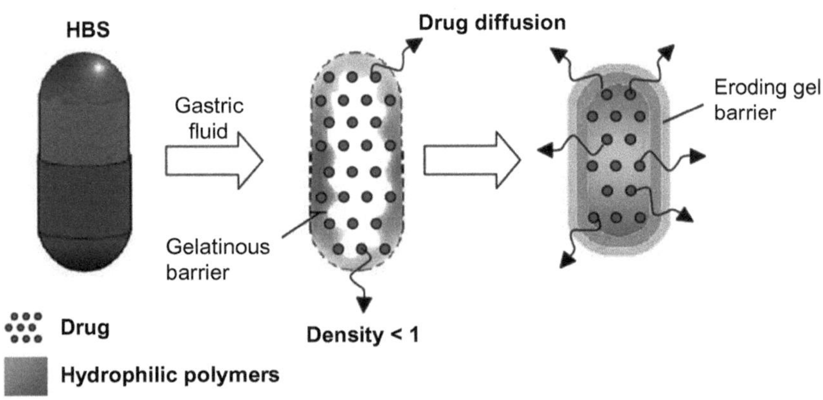

Figure 6.37 Hydrodynamically balanced system (HBS): The gelatinous polymer barrier formation results from hydrophilic polymer swelling; the active agent is released by diffusion and erosion of the gel barrier.
Reprinted from Bardonnet et al. (2006), with permission from Elsevier.

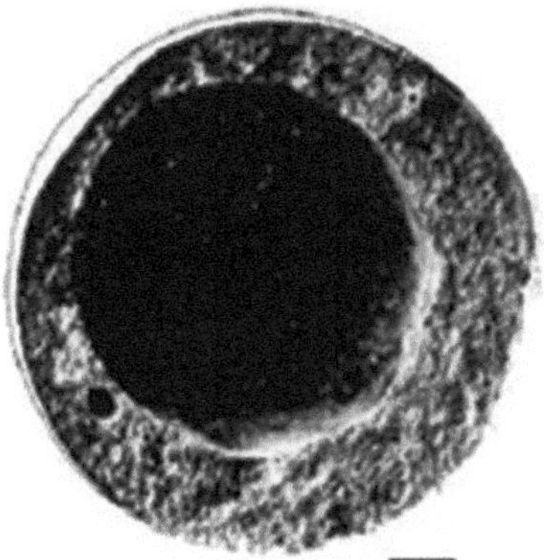

Figure 6.38 Micrograph of a hollow microsphere (or microballoon).
Reprinted from Bardonnet et al. (2006), with permission from Elsevier.

reactions that produce carbonic gas can be cited. All of them aim to obtain a dosage form with lower density than gastric fluid.

Examples of multiparticle systems based on hollow microcapsules containing a drug in the membrane include Eudragit® S with ibuprofen and polycarbonate with acetylsalicylic acid. These microcapsules are prepared using a modified method from the classic method of emulsification and solvent evaporation. To obtain the floating microspheres, the drug is dissolved in a polymeric solution of volatile organic solvent. This solution is emulsified in an aqueous phase containing a tensoactive, normally polyvinyl alcohol. After the solvent evaporation, the microparticles display a polymeric membrane plus an active agent involving an aqueous core. After drying, the microparticles lose the inner water and have the ability to float in the gastric fluid.

The obtainment of microcapsules (reservoir systems) and not microspheres (matrix systems) is dependent on the conditions used during preparation.

Another type of floating system uses porous particles of calcium silicate as a floating unit. The pores are closed with ethyl cellulose or HEC. The floating is due to the lower density of captive air into the pores of particles. The viscosity and superficial tension of the coating liquid induces the formation of an impermeable membrane covering the opening core. This membrane does not involve the inner wall of the pore. Controlling the amount of coating material, it is possible to close part of the pores, and those not closed can be filled with the active agent.

Ichikawa, Watanabe, and Miyake (1991) developed a multilayer dosage form by coating using a fluidized bed. The active agent (*p*-aminobenzoic acid) is contained in the core and it is coated with hydroxypropyl cellulose. The next coating is composed of a layer of sodium carbonate and the other of tartaric acid, separated by a layer of HEC. Externally, there is a swelling and elastic coating of hydroxypropylmethylcellulose phthalate and polyvinyl acetate. After the administration, the system absorbs water, which causes the dissolution of tartaric acid and sodium bicarbonate. They react, forming carbonic gas, which expands without rupturing the external membrane. The system becomes less dense and floats.

6.9.4 Applications

Gastro-retention is essential for active agents that are absorbed from the stomach, those that are poorly soluble or degraded by the higher pH of the intestine, and active agents with an absorption that can be modified by changes in gastric emptying time.

Therefore, floating systems are useful for sustained drug delivery and site-specific drug delivery, and to enhance the absorption of the active agent. For example, local as well as sustained delivery of the active agent or certain conditions (e.g., *Helycobacter pylori* infection, which is the cause of peptic ulcers) may be cited. This strategy can improve bioavailability and therapeutic efficacy, and may even also allow a possible reduction in the dose because of steady therapeutic levels of the active (e.g., ofloxacin and furosemide). The reduction in fluctuations in therapeutic levels minimizes the risk of resistance, especially in the case of β-lactam antibiotics (penicillins and cephalosporins).

Molecules with poor colonic absorption that are characterized by better absorption at the upper sites of the gastrointestinal site are, generally, the appropriate candidates for floating drug delivery systems. Examples are active agents that display a narrow absorption window in the gastrointestinal tract (e.g., levodopa and riboflavin), active agents that are primarily absorbed from the stomach and upper part of the gastrointestinal tract (e.g., calcium supplements, chlordiazepoxide, and cinnarizine), active agents that act locally in the stomach (e.g., antacids), active agents that degrade in the colon (e.g., hydrochloride of ranitidine and metronidazole) and active agents that disturb normal colonic bacteria (e.g., trihydrate of amoxicillin).

6.9.5 Limitations

Like all other drug delivery systems, floating systems have some limitations. Active agents that are unstable in the acidic environment of the stomach are not suitable candidates to be incorporated into the systems. Actives that are irritants to gastric mucosa are also not desirable or suitable. Active agents that are significantly absorbed throughout the gastrointestinal tract, which undergo significant first-pass metabolism, are the only desirable candidates.

6.9.6 Future and perspectives

Oral administration of active agents is conventional and well accepted by patients, improving treatment. However, during the development of an efficient floating gastro-retentive drug delivery system, it is necessary to consider that the system must remain for a sufficient time in the stomach, which is not compatible with its normal physiology.

As previously discussed, gastro floating systems display potential as a future approach for gastric retention. Due to their limited absorption in the upper gastrointestinal tract, active agents displaying poor bioavailability can be released efficiently, thereby maximizing their absorption and improving their absolute bioavailability.

6.10 Inclusion complex with cyclodextrins

Cyclodextrins are macrocyclic oligosaccharides linked by α-1,4 glycosidic bonds with important interest in supramolecular chemistry. They have annular structure and the ability to produce inclusion complexes, which represent true encapsulations at the molecular level.

Research and use of these complexes has increased in recent decades. Their importance is so big that "it is difficult to imagine a world without cyclodextrins" (Kurkov & Loftsson, 2013).

The first studies about cyclodextrins began in the late nineteenth century with the observations of French scientist Villiers. He observed crystals dispersed in alcohol waste left after the production of dextrins from starch, using an impure bacterial culture (Villiers, 1891). The chemical composition and some chemical properties of these crystals were determined and, during the twentieth century, various researches were performed to characterize the cyclodextrins as well as elucidate their mechanism of production by different microorganisms (Loftsson & Duchêne, 2007).

6.10.1 Classification

Cyclodextrins are classified according to the numbers of glucopyranose units they have in their structure. Therefore, when the molecule of cyclodextrin has six glucopyranose units, the denomination is α-cyclodextrin. For seven and eight units of glucopyranose, they are denominated β-cyclodextrin and γ-cyclodextrin, respectively.

Due to the conformation of glucopyranose units, bound via α-1,4-glycosidic linkages, these supramolecular structures have incredible features. The inner space of the cyclodextrin molecule is hydrophobic, while the exterior is composed of hydroxyl groups, displaying polar or hydrophilic behavior. This characteristic allows the formation of inclusion complexes with molecules or groups of the same molecules displaying different polarities. The truncated cone-shaped cyclodextrins possess a hollow, tapered cavity of 0.78 nm in depth, while both the top and bottom diameters are increased with the number of glucose units (Figure 6.39).

As soon as the cyclic structure of cyclodextrin molecules was disclosed, it was proposed that they should be able to include molecules in their cavity. Lipophilic molecules can fit into the ring to form soluble inclusion complexes.

Due to the hydrophobic cavity, cyclodextrins have inclusion capacity with a variety of substances, ranging from ions to small molecules, oligonucleotides and proteins. β-cyclodextrin possesses the ring with the correct size for various molecules of active agents. Different stoichiometries are possible, but normally one molecule of the active agent associates with one cyclodextrin molecule to form reversible complexes.

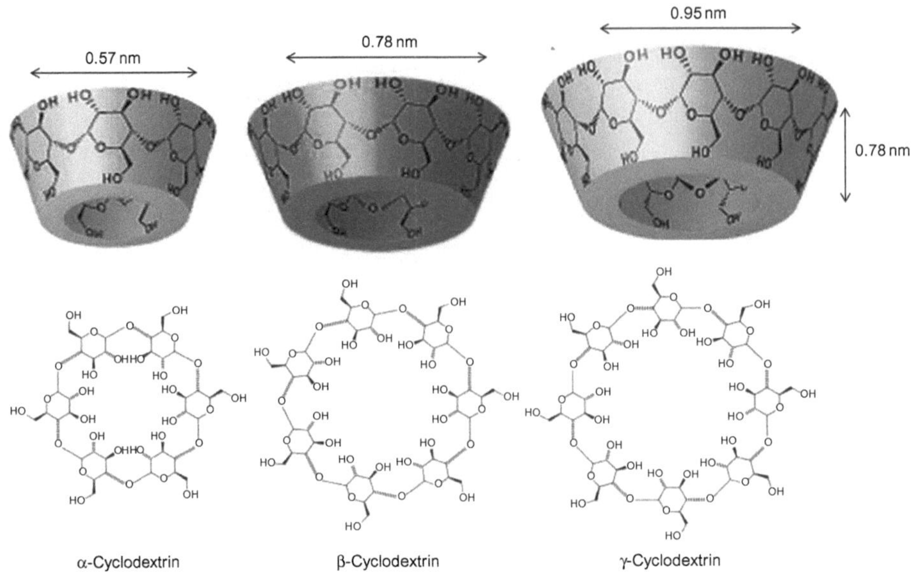

Figure 6.39 Structure and inner dimensions of cyclodextrins.

Its structural characteristics predetermined the application of cyclodextrin as a solubilizer for poorly water-soluble chemicals and as a modifier of drug delivery. Miconazole, itraconazole, piroxicam, indomethacin, digitoxin, naproxen, hydrocortisone, diazepam and pilocarpine are examples of drugs that improved their bioavailability due to enhancement of solubility.

Aiming to improve the pharmaceutical features of native cyclodextrins (solubility, inclusion capability, controlled drug delivery capacity and reduction of toxicity), some semi-synthetic compounds have been obtained. For this, chemical modifications are used to incorporate methyl, ethyl, hydroxyethyl, carboxymethyl and hydroxypropyl groups to the original structures, giving derivatives with different solubilities (highly soluble, amphiphilic and hydrophobic derivatives). Nowadays, more than 30 types are marketed as pharmaceutical products.

Amphiphilic cyclodextrins have increased ability to interact with biological membranes, rendering self-assembly capacity in aqueous solutions. They can be classified into neutral, cationic, and anionic.

Hydrophobically modified cyclodextrins have been used to develop sustained-release systems. Alkylated, acylated and acetalated derivatives have been proposed as cyclodextrin derivatives.

Moreover, cyclodextrin-containing polymers have been synthesized to obtain materials with improved features. Using polymers of various structures, it is possible to prepare cyclodextrin-containing polymers that display multiple recognition sites for molecular self-assembly, enhancement of biocompatibility of polymers for biomedical applications, production of functional materials for controlled delivery of active agents and gene therapy. These improved features and applications are possible due to the diverse architectures of polymers (linear, grafted, block, branched, hyperbranched, and dendritic) and the possibility to covalently link the cyclodextrin in main chains or conjugated as flanking side groups (Zhang & Ma, 2013).

Cyclodextrin-based polypseudorotaxanes are a family of supramolecular polymers containing cyclodextrin threading onto the polymer chains, providing abundant hydroxyls in the resulting materials. Therefore, many active agents may be conjugated on the polymers to produce supramolecular prodrugs. In addition, recognizable units (e.g., carbohydrate ligands, peptides, proteins, apatamers or other nucleic acids) can be added to the structure to result in additional functions, such as targeting to specific molecules, cells or tissues.

Considering their important clinical performance for tumor therapy and diagnosis of cardiovascular diseases, nanoscaled supramolecular systems have been considered the most promising carriers for drug and gene delivery as well as other biomedical applications. They can prolong circulation and they have broad payload spectrum, unique size and shape characteristics for tissue penetration and passive targeting, as well as specific cellular/subcellular trafficking pathways. In addition, they are easy to tailor for active targeting at different levels and facile control of therapeutic release by sophisticated material engineering (Zhang & Ma, 2013).

The introduction of cyclodextrin into supramolecular platforms may result in enhanced biocompatibility, functionalizing flexibility, and recognition capability for both the active agent payload and the therapeutic target.

In this sense, vesicles, micelles, nanoparticles, nanogels, nanopolyplexes, supramolecular hydrogels, highly complex superstructures and multifunctional systems can be produced by the strategy of using cyclodextrins and polymeric materials.

6.10.2 Drug release from cyclodextrin complexes

Cyclodextrin complexes are used for drug release, considering that their main property is the possibility of simple dissolution of a complexed active agent with cyclodextrin and dilution in aqueous media.

Inclusion complexes are stabilized by intermolecular forces such as van der Waals and hydrogen bond. Molecular encapsulation can occur in the solid state or in solution.

Processes such as drug–protein binding, direct drug partitioning from the complex to tissue and competitive binding contribute to rapid drug release from the complexes (Kurkov & Loftsson, 2013).

Cyclodextrins increase the oral absorption of active agents and, with only a few exceptions, the cyclodextrin-active agent complexes do not hamper their therapeutic effect. It has been shown that the binding constant of drug-cyclodextrin complexes must be greater than approximately 10^5/M to have any effect on the active pharmacokinetics after parenteral administration (Stella & He, 2008).

Therefore, the pharmaceutical uses of cyclodextrin complexes aim to increase the hydrosolubility of drugs, reduce their adverse affects and increase their stability. Moreover, the combination of incompatible drugs, taste and odor masking, and the transformation of liquid active agents in solids are possible.

6.10.3 Applications

Cyclodextrin is used in drug delivery systems as a strategy to modify the release of therapeutics, including formulations intended for parenteral administration. Nowadays, cyclodextrins are being highly utilized due to the technological advances in their production, which have reduced their costs.

Cyclodextrins have various applications:

(1) Increasing drug solubility and stability
(2) Enhancing drug absorption
(3) Masking odors and tastes
(4) Controlling drug release profiles
(5) Alleviating local and systemic toxicity
(6) Improving drug permeability across biological barriers

The inclusion complex using cyclodextrin can increase the stability of a drug molecule (e.g., against the heat), leading to the reduction of volatility or the increase of thermal resistance.

Cyclodextrins and their derivatives have been successfully employed to construct supramolecular systems across length scales and to engineer novel functional materials, taking full advantage of host–guest interactions between the cyclodextrin units and guest molecules. In this context, cyclodextrins can be utilized as molecular valves to switch the ON/OFF release of payload from hybrid nanosystems (Zhang & Ma, 2013).

A great variety of cyclodextrin-containing polymers with versatile architectures have been synthesized to assemble functional platforms. These compounds possess important applications in gene therapy, drug delivery, and medical imaging.

6.10.4 Advantages and disadvantages

Cyclodextrins are nontoxic compounds with the ability to complex and stabilize a large range of substances. Due to their low toxicity and low immunogenicity, cyclodextrins have extremely attractive pharmaceutical applications.

They are produced from the starch by reaction of cyclization of linear chains of glucopyranose by cyclodextrin glucosyl transferase. Normally, this enzyme produces a mixture of γ-, β-, and γ-cyclodextrins, and it is necessary to separate and purify them.

Pharmaceutical formulations containing cyclodextrins can be dosed by oral, nasal, ocular, rectal, and dermal delivery.

The parenteral administration of cyclodextrin can cause nefrotoxicity, particularly the β-cyclodextrin, due to the formation of low-solubility complex β-cyclodextrin/cholesterol, which precipitates in the kidneys. Parenteral administration can also cause hemolysis; therefore, derivative substances of cyclodextrins were developed. They have the same properties, but are much more soluble (e.g., hydroxypropyl-β-cyclodextrin).

6.11 Transdermal drug delivery systems

The human skin is a readily accessible surface for drug delivery. Topical application has been used for centuries, predominantly in the treatment of localized skin diseases. Local treatment requires only that the active agent permeate the outer layers of the skin to treat the diseased state, with little or no systemic accumulation. However, medicines that are applied to the skin can release active agents afterward to cross the various skin layers and reach the blood system.

Therefore, transdermal drug delivery is distinct from topical drug administration, which targets local areas. In the last three decades, transdermal drug delivery has become a viable way of delivering a number of active agents, with the potential to deliver many more.

Transdermal systems are designed to obtain systemic blood levels. They represent a technological evolution of the classic patch.

With the understanding of the principal permeation barrier in the skin, it was possible to understand that the drug penetration through the skin can occur simultaneously by both transcellular and intercellular pathways. Moreover, it is possible that drug penetration bypassing the skin appendices (glandules and bristles) occurs, called transappendicular penetration. This transdermal permeation, or percutaneous absorption, is defined as the passage of a substance (e.g., an active agent) from the outside of the skin into the bloodstream.

Considerable efforts have been expended to design formulations or devices to deliver enough of the active agent such that there is a sufficient amount present at the site of action.

6.11.1 Advantages and disadvantages

Over the last seventy years, transdermal drug delivery has been studied quite a bit, with great expectations about the blood levels controlled by the transdermal device.

For many reasons, this strategy is not considered universal for the administration of medicines. However, this route for drug delivery remains attractive and offers many benefits.

Considering that the skin itself controls drug input into the blood circulation, drug delivery can be controlled predictably, and over a long period of time. Therefore, transdermal systems display some potential advantages:

(a) Avoid the hostile environment of the gastrointestinal tract.
(b) Avoid physiological variations or other contraindications of oral delivery.
(c) Enable suitable absorption of the active agent.
(d) Increase patient compliance.
(e) Avoid the first-pass effect.
(f) Allow the effective use of drugs with short biological half-lives.
(g) Allow the administration of drugs with narrow therapeutic windows.
(h) Provide controlled plasma levels of highly potent active agents.
(i) Drug delivery can be promptly interrupted in the case of toxicity.

On the other hand, considering that the skin is a very effective barrier for the permeation of most xenobiotics, an active agent that is incorporated into a transdermal drug delivery system will exhibit low bioavailability. Therefore, these systems show some drawbacks:

(a) It is impossible to administer active agents that need high blood levels.
(b) The adhesive layer cannot adhere well to all types of skin.
(c) The active agent or formulation can irritate the skin or produce sensitization.
(d) The system may not be comfortable to wear.
(e) Sometimes, the system is expensive.

Transdermal administration offers a non-invasive and convenient method for pediatric drug delivery. However, the immature and rapidly evolving skin barrier function in premature neonates represents a significant formulation challenge. Moreover, the significant risk of excessive drug exposure via the incompletely formed skin barrier needs to be considered.

6.11.2 Classification

The design of an effective transdermal delivery system involves approaches to acquire high enough permeability of the active agent through the stratum corneum, ensuring that the delivery system did not irritate the skin, the active agent was not unduly metabolized, the active delivered in this manner had appropriate pharmacokinetics and pharmacodynamics, and the drug was stable in the system (Langer, 2004). In this sense, skin characteristics, bioactivity of the active agent, adhesion, formulation and design of the device must be taken into consideration during the development of transdermal delivery systems.

Transdermal passive systems are therapeutic systems in which drug release is factory-set. Thus, each device has a drug amount higher than necessary, or contains devices (membranes or matrices) that allow the creation of a gradient of concentration between the dosage form and the inner layers of skin. This gradient enables the drug diffusion to the epidermis at a constant rate during a predetermined time.

Among the many strategies and devices proposed for transdermal drug delivery, skin patch devices are the most common and probably the most innovative practical step in the science of transdermal delivery in recent years.

Transdermal therapeutic systems (TTS) were introduced as devices displaying the ability to release an active agent to the skin in a controlled manner, well below the maximum the tissue can accept. The principle is that the system would control the rate at which an active agent diffuses through the skin. Therefore, the intended flux would be much lower than the maximum skin flux.

In this context, patches were developed in a variety of ways, but classified into some main types: reservoir type, micro reservoir type, matrix type, membrane matrix hybrid, and active agent in adhesive type. Some examples are illustrated in Figure 6.40.

6.11.2.1 Reservoir type

This type of transdermal patch contains a drug reservoir made of a homogenous dispersion of an active agent suspended in an unleachable viscous liquid medium (e.g., silicon fluids), forming a paste-like suspension or gel or a clear solution of the active in a releasable solvent such as ethanol.

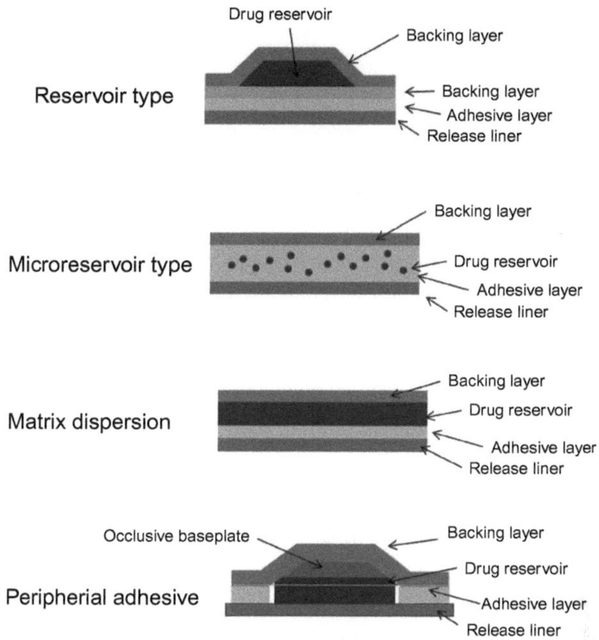

Figure 6.40 Schematic designs of transdermal drug delivery patches.

The reservoir is placed between a rate-controlling membrane and backing laminate. The release rate is controlled by the diffusion of the active agent directly through the material, or the material may contain fluid-filled micropores, in which case the active may additionally diffuse through the fluid.

As discussed in previous chapters, in the case of nonporous membranes, the rate of transport of drug molecules depends on their solubility in the membrane and the thickness of the membrane. The main materials used as membranes are ethyl cellulose, silicon rubber, ethylene vinyl acetate (EVA), and polyurethanes. These enable the production of patches, which can provide a true zero-order release pattern to achieve a constant serum drug level.

6.11.2.2 Micro reservoir type

Unleachable microscopic drug reservoirs (micro reservoirs) are obtained by dispersion of the solid active agent in an aqueous solution of a water-miscible drug solubilizer, such as polyethylene glycol. The dispersion is submitted to high-shear mechanical force in a hydrophobic polymer.

The tendency of the micro reservoirs is to joint. Therefore, the polymer is cross-linking *in situ* to stabilize the dispersion and to produce a medicated polymer disc of a specific area and fixed thickness.

The loss of the active agent through the backing membrane is prevented by the use of an occlusive baseplate mounted between the medicated disc and adhesive form backing.

6.11.2.3 Matrix type

The active agent and the polymer are dispersed together, and the required quantity of plasticizer (e.g., dibutylpthalate, triethylcitrate, polyethylene glycol, or propylene glycol) and permeation enhancer are added and mixed properly. The matrix with the active agent is then molded into rings with defined surface area and controlled thickness over the mercury on a horizontal surface, followed by solvent evaporation at an elevated temperature.

The result is the film, which is separated from the rings and mounted onto an occlusive baseplate in a compartment fabricated from an impermeable backing to the active agent. Adhesive polymer is then spread along the circumference of the film.

For example, some matrix materials used are cross-linked polyethylene glycol, Eudragits, ethyl cellulose, polyvinylpyrrolidone, and hydroxypropylmethylcellulose.

6.11.2.4 Membrane matrix hybrid type

This type is a modification of the reservoir-type transdermal patch. The liquid formulation of the reservoir is replaced with a solid polymer matrix like polyisobutylene, which is placed between the rate-controlling membrane and backing laminate.

6.11.2.5 Drug in adhesive type

Another type of patch is when the active agent is dispersed in the adhesive material. The active and other selected excipients are directly incorporated into the organic solvent-based pressure-sensitive adhesive solution, mixed, cast as a thin film and dried to evaporate the solvents, leaving a dried adhesive matrix film containing the drug and excipients. This mixture is placed between the release liner and backing layer. This type of patch can be single layer or multilayer. The multilayer system is different from single layer in that it adds another layer of drug-in-adhesive, usually separated by a membrane.

Despite the differences between the types, the main components of a transdermal patch are the polymer matrix or membrane, active agent, permeation enhancers, adhesive, backing laminates, release liner, and other excipients (e.g., plasticizers and solvents).

When the system is a matrix, the polymer matrix is its backbone and controls the release of the active agent. On the other hand, when the system is a reservoir, there exists a polymer membrane with the function to control the release of the active agent. The polymers used in these cases should be low-cost, nontoxic and chemically nonreactive, and should not decompose on storage. Examples include zein, gelatin, waxes, cellulose derivatives, gums, polybutadiene, hydrin rubber, polyisobutylene, silicon rubber, nitrile, acrylonitrile, neoprene, polyvinyl alcohol, polyvinyl chloride, polyethylene, polypropylene, polyacrylate, polyamide, polyurea, polyvinylpyrrolidone, and polymethylmethacrylate.

The active agents to be delivered by transdermal patches must be substances that undergo extensive first-pass metabolism, those with a narrow therapeutic window, or actives with a short half-life.

To increase the stratum corneum permeability, so as to attain higher therapeutic levels of the active agent, many times it is necessary to use permeation enhancers. Another component of TTS is the adhesive, which has the function to adhere the device to the skin.

Moreover, backing laminates are highly flexible materials (e.g., vinyl, polyethylene) that constitute the back of the device. The release liner protects the patch during storage and needs to be removed prior to use.

Other excipients are used in the preparation of patches. Acetone, methanol, chloroform, isopropanol and dichloromethane are used to prepare drug reservoirs. Moreover, plasticizers (e.g., dibutylpthalate, triethylcitrate, polyethylene glycol, and propylene glycol) are added to provide plasticity to the transdermal patch.

6.11.3 Applications

Modern transdermal patch medicines can be traced back to 1979, when scopolamine patches were approved by the US Food and Drug Administration (FDA). Nowadays, numerous transdermal patches for active agents are available in the market (e.g., nitroglycerin, nicotine, scopolamine, clonidine, fentanyl, estradiol, testosterone, lidocaine, and oxybutinin). Depending on the active, the time of delivery duration is generally from 1 to 7 days. A very important example of a successful system on the market is the estradiol patch, which is used by more than a million patients worldwide per year. In contrast to oral formulations, these are not associated with liver damage.

Moreover, transdermal clonidine, nitroglycerin, and fentanyl patches exhibit fewer adverse effects than conventional oral dosage forms (Langer, 2004). Nicotine patches have also played an important role in preventing smoking and prolonging life.

6.11.4 Transdermal permeation enhancement and strategies

When a substance reaches intact skin, there are three potential entry pathways to the viable tissue: across the continuous stratum corneum, through the hair follicles with their associated sebaceous glands and via the sweat ducts. Some biological factors have influence on transdermal permeation (e.g., skin condition, skin age, blood flow, and skin metabolism) as well as physicochemical factors (e.g., skin hydration, temperature, pH, diffusion coefficient, drug concentration, partition coefficient, molecular size, and shape).

Most active agents do not permeate the skin in therapeutic amounts. Substances that are administered across the skin are of low molecular mass (<500 Da) and are very lipophilic in nature at low dosages. Hydrophilic substances generally exhibit poor skin permeability (10^{-7} to 10^{-8} cm/s), about one or more orders of magnitude lower than hydrophobic solutes (Azagury, Khoury, Enden, & Kost, 2014).

The skin barrier and its relative impermeability led to the development of a number of enhancements to chemical and physical approaches.

The chemical approaches comprise methods that produce the active at very high thermodynamic activity.

One chemical method is based on solvent loss, which may occur as a result of its evaporation or its diffusion into the skin. Moreover, the system matrix may take up water from the skin, leading to altering its solubility properties such that the permeant becomes supersaturated. The close-to-saturation state of the active agent can cause stability issues, and permeation rates from systems change with time.

It is common to use solvents that permeate into the skin and act essentially as a carrier for the active agent. These solvents can be ethanol, dimethyl sulfoxide, propylene glycol, glycerin, polyethylene glycol, urea, dimethyl acetamide, sodium lauryl sulfate, the poloxamers, Spans, Tweens, lecithin, terpenes, and many others, which can co-diffuse with the active agent, enabling its passage through the stratum corneum.

Another chemical approach is based on the presence of a component (e.g., surfactant-like substance) that permeates into the intercellular lipids, where it intercalates and disrupts their structure. This action led to creating a region where diffusion is faster and permeation through the stratum corneum is improved. These enhancers can have irritant properties and need to be selected considering the reversibility of barrier function affection. It is very important that they do not disrupt the membranes of the viable cells in the deeper layers of the skin, where they could elicit adverse effects. They are divided into chemical groups such as sulfoxides, pyrrolidones, fatty acids, alcohols, surfactants, metabolic interventions, and the only material specifically designed to enhance transdermal mass transport, Azone (Williams & Barry, 2004).

These three chemical approaches can be used simultaneously, so that in a combination of enhancers, one acts as a good solvent and one that disrupts the lipid structure.

Several physical strategies for skin penetration enhancement, like stripping of the stratum corneum, heating, iontophoresis, electroporation, ultrasound, and microneedles, have also been studied.

With the high advance in the development of biotechnological products, and the need to deliver peptides, oligonucleotides and their analogs, it is difficult to use conventional chemical enhancer methods. In addition, most peptides are charged and contribute to the iontophoresis approach to be used.

Iontophoresis is a technique of the nineteenth century, by which the charged active agent moves through the skin membrane using an electrical field. It involves the application of a small electric current (up to $0.5\,mA/cm^2$) to a drug reservoir, wetting the surface of the skin with the same charged electrode as the solute of interest. The process is dominated by the physicochemical properties of the active agent and considers the pH of the stratum corneum (4–5 to physiological). If the pKa of the permeant is between these values, charge reversal can occur, and the potential difference drives the active agent into the skin. Moreover, due to viscous forces, the entire solution undergoes a convective flow that carries non-charged particles as well. This process is referred to as electro-osmosis and can facilitate the absorption of uncharged hydrophilic substances. A number of actives have been submitted to iontophoretic delivery, such as lidocaine, amino acids, peptides, proteins, dexamethasone, verapamil, and propranolol.

Electroporation involves the application of a pulsating electrical field at high voltage ($>50\,V$, typically 1–100 ms) to the skin, causing the formation of transient aqueous pores in the stratum corneum, through which molecular transport is attainable. This technology was originally used to transfect cells with macromolecules such as DNA.

In the skin, electroporation leads to enhanced permeability, mainly attributed to electrophoretic movement and diffusion through the newly created aqueous pathways. To differentiate from iontophoresis, which directly acts on the drug molecule to propel it into the skin, electroporation acts mainly on the skin to increase its permeability.

Transdermal delivery mediated by ultrasound is also named phonophoresis or sonophoresis. Sonophoresis has relatively higher cavitation and improves the transport of substances through the skin and into the soft tissue during or following the influence of an ultrasonic perturbation (20 kHz–1 MHz) (Azagury et al., 2014). The fundamental mechanism by which sonophoresis enables transdermal delivery is still not clearly understood or characterized. There are proposed mechanisms such as thermal effects by absorption of ultrasound energy and cavitation effects caused by collapse and oscillation of cavitation bubbles in the ultrasound field. Between these two effects, cavitation is believed to be the predominant mechanism responsible for sonophoresis.

The barrier function of the skin can also be physically breached by "shooting" particles through it using high velocities or by ablating the stratum corneum by precisely controlled laser technology. Laser cell-ablation to remove the stratum corneum barrier has been investigated as a means of enhancing transdermal drug delivery. Lasers such as erbium—yttrium–aluminum–garnet (YAG)—have been found to increase skin permeability. Molecular size, lipophilicity and sequence of peptides were found to play important roles in modulating delivery enhancement.

Radio-frequency cell ablation is a physical approach by which cell ablation is performed by placing an array of microelectrodes on the skin and passing an alternating electrical current at radio frequency (100–500 kHz). The ions in the cells adjacent to the microelectrodes vibrate as they try to follow the change in electrical current direction. These vibrations generate heat, which causes water evaporation, cell ablation and possibly damage of deeper skin layers (Azagury et al., 2014). The microchannels are formed by placing a closely spaced array of tiny electrodes with very precise dimensions against the skin. The alternating electrical current transfers through each of the microelectrodes, ablates the cells underneath each electrode and forms microscopic passages in the stratum corneum and in the outer dermis.

Microneedles can also be developed to pierce the skin to allow delivery to carefully controlled depths. They create a physical pathway through the upper epidermis to increase skin permeability. The device is designed containing some microneedles that, when applied to the skin surface, pierce the outer epidermis layer deep enough to increase skin permeability and allow delivery of the active agent (Figure 6.41). Because the microneedles reach only sites that contain no nerves, they do not cause any pain through the sensory receptors of the dermis. The main materials used to prepare microneedles are silicon and biodegradable materials such as carboxymethyl cellulose, solid maltose, polylactic acid (PLA), poly(lactic-co-glycolic acid) (PLGA), and polyvinylpyrrolidone (Figure 6.42).

6.11.5 Future and perspectives

Skin represents a very important pathway to drug delivery, in that it can provide an effective means for delivering drugs that are destroyed by the liver when taken orally.

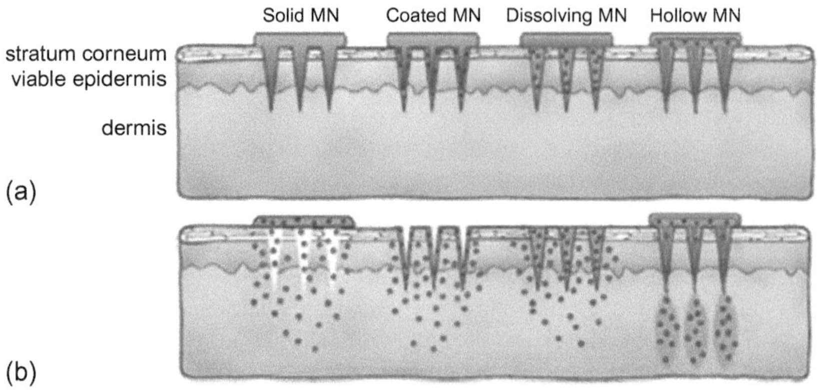

Figure 6.41 Methods of drug delivery to the skin using microneedles (MN). Microneedles are first applied to the skin (a) and then used for drug delivery (b). Solid microneedles are used as a pretreatment, after which the drug can diffuse through residual holes in the skin from a topical formulation (solid MN). After insertion of drug-coated microneedles into the skin, the drug coating dissolves off the microneedles in the aqueous environment of the skin (coated MN). Drug-loaded microneedles are made of water-soluble or biodegradable materials encapsulating the drug that is released in the skin upon microneedle dissolution (dissolving MN). Hollow microneedles are used to inject liquid formulations into the skin (hollow MN). Reprinted from Kim, Park, and Prausnitz (2012), with permission from Elsevier.

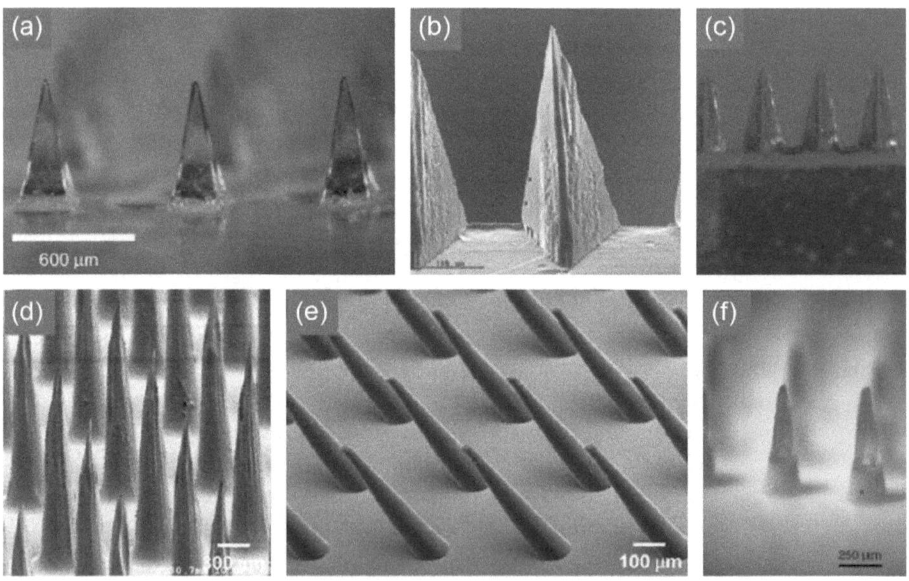

Figure 6.42 Scanning electron microcopy and bright-field microscopy pictures of biodegradable microneedles of (a) carboxymethyl cellulose; (b) solid maltose; (c) solid maltose; (d) PLA; (e) PLGA; (f) PVP.
Reprinted from Ochoa, Mousoulis, and Ziaie (2012), with permission from Elsevier.

Transdermal studies are focusing on better transdermal device design and examining how different enhancers interact with each other. With better understanding of skin irritation, immunology and metabolism, the development of more effective transdermal systems for delivery of a wide variety of molecules will be possible.

However, it is necessary to consider that the development of those medicines must be less expensive. Therefore, the aim is to make active materials more active, selective and delivered in a controlled manner. Transdermal delivery is a strategy that shows that the relative costs of using a complex sophisticated delivery system will become less of a problem in the future.

This strategy and its approaches are enabling the administration of a far wider range of active agents that otherwise could not be delivered using this attractive route of administration. The non-invasive monitoring of drug levels or of other biological markers is becoming a reality and, therefore, it should be possible to program drug delivery accurately and only when required (Azagury et al., 2014).

6.12 Bioadhesive systems

Bioadhesion can be defined as the state in which two biological materials, or a biological and a synthetic material, are maintained together for a prolonged time period by means of interfacial forces (Smart, 2005).

Over the last 30 years, this strategy began to be applied to drug delivery systems (Bruschi & de Freitas, 2005; Carvalho, Bruschi, Evangelista, & Gremião, 2010). In this context, the term bioadhesion is utilized to describe the adhesion between materials (natural or synthetic) and biological soft tissues (e.g., epithelial cells and mucus layer). When there is the presence of mucous tissue, the bioadhesion is named mucoadhesion.

Therefore, the term bioadhesion is inclusive and used to describe adhesive interactions with any biological or biologically derived substance. In addition, the term mucoadhesion describes the interaction between the systems and mucus or mucous tissue.

Bio/mucoadhesive drug delivery systems are obtained by the incorporation of bioadhesive molecules into some kind of pharmaceutical system intended to stay in close contact with the absorption tissue, releasing the drug near the action site, thereby increasing its bioavailability and promoting local or systemic effects (Hägerström, Edsman, & Strømme, 2003; Woodley, 2001).

This approach constitutes a very important strategy to prolong the residence time at the absorption site, allowing intensified contact with the adhesion of preparations onto mucous or epithelial tissue. Therefore, it is possible to retain the preparation at the action site and to direct the active agent to a specific site or tissue. Other features associated with the development of controlled drug delivery systems using bioadhesive molecules include a decrease in drug administration frequency and an increase in patient compliance to the therapy (Woodley, 2001).

The bioadhesive process is dependent on the presence of bonds between the two surfaces. It is very important to understand the forces involved in this process of bond formation. Moreover, it is necessary to consider that different types of biological surfaces such as mucosal membranes and epithelial cells may be present. Thus, the process of bond formation is dependent on the wetting and swelling of polymeric material; the interpenetration of bioadhesive polymer chains, conducing to entanglement with the mucin chains; and the formation of weak chemical bonds between the entangled chains (Carvalho et al., 2010).

For this process, the polymeric material needs to have sufficient amounts of hydrogen-bonding chemical groups (–OH and COOH), surface charges, high molecular weight, high chain flexibility and surface tensions that will spread into the mucus layer (Peppas & Buri, 1985). Chemical bonds are strong primary bonds, such as covalent bonds, and weaker secondary forces (e.g., ionic bonds, van der Waals interactions, and hydrogen bonds). Mechanical interactions can be useful to make connections between surfaces, involving the inclusion of a substance in the crevices of another. Microscopically, the interpenetration of mucin strands into a porous polymer surface and/or physical entanglement of mucin strands with flexible polymer chains can occur. These processes are dependent on the chain flexibility and diffusion coefficients of each material.

6.12.1 Mechanism of bioadhesion

Many theories have been proposed to explain bioadhesion. The electronic theory explains bioadhesion by the assumption that the bioadhesive material and the biological site have different electronic structures. In this sense, when the materials come into contact, electron transfer occurs and a double layer of electrical charge forms at the interface. The bioadhesive force is due to attractive forces across this electrical double layer.

The adsorption theory explains bioadhesion by the bond formed between the surfaces due to the van der Waals interactions, hydrogen bonds and related forces. Despite the weak characteristics of these forces, the sheer number of interactions can as a whole produce intense adhesive strength.

The wetting theory explains bioadhesion by the ability of two surfaces (e.g., polymer and mucus) to spread and develop intimate contact with their corresponding substrate. This theory is based on the capacity of materials to spread. Therefore, it is very important to study and consider the surface energy of both materials and tissues to predict bioadhesive performance.

The diffusion theory explains bioadhesion by the interpenetration and entanglement of bioadhesive material chains and mucous chains to produce semi-permanent adhesive bonds. It is clear that bond strength increases with the degree of penetration of the polymer chains into the mucus layer.

The fracture theory is based on analysis of forces required to separate two surfaces after adhesion. The maximum tensile force produced during the detachment divided by the total surface area involved in the adhesive interaction results in the force of bioadhesion.

These theories are used together to understand and explain bioadhesion.

Despite the need for bioadhesive formulations for dermal administration when prolonged cutaneous action is desired, the mucous membrane is the main administration site for bioadhesive systems (Carvalho et al., 2010).

In this context, human mucous membranes are characterized by an epithelial layer covered by mucus, showing relative permeability and allowing fast absorption of active agents. Mucus has the important function of epithelium protection and lubrication, being composed of glycoproteins, lipids, inorganic salts and 95% water by mass, making it a highly hydrated system. The most important glycoprotein of mucus is mucin, which is responsible for mucus structure. The thickness of the mucus layer varies from 50 to 450 μm in the stomach to less than 1 μm in the oral cavity (Smart, 2005).

Considering the bioadhesion theories, the mechanism of mucoadhesion is generally divided into two stages (Figure 6.43): the contact stage and the consolidation stage.

The first stage is characterized by contact between the mucoadhesive and the mucous membrane, with spreading and swelling of the formulation, initiating its deep contact with the mucus layer. In some cases, such as for ocular or vaginal formulations, the delivery system is mechanically attached over the membrane (Pereira, Godoy, Svidzinski, & Bruschi, 2013).

In other cases, the deposition is promoted by the aerodynamics of the organ to which the system is administered, such as for the nasal route. On the other hand, in the gastrointestinal tract, direct formulation attachment over the mucous membrane is not feasible. Some physiological phenomena can contribute to mucoadhesion, like peristaltic motions that contribute to the contact. However, undesirable adhesion in the esophagus can occur. In these cases, mucoadhesion can be explained by peristalsis, the motion of organic fluids in the organ cavity, or by Brownian motion (Carvalho et al., 2010).

During the consolidation stage, the mucoadhesive materials are activated by the presence of moisture, which plasticizes the system, allowing the mucoadhesive molecules to break free and to link up by weak van der Waals and hydrogen bonds (Smart, 2005).

The diffusion theory and the dehydration theory explain the consolidation step (Figure 6.44).

The mucoadhesive molecules and the glycoproteins of the mucus mutually interact by means of interpenetration of their chains and the building of secondary bonds. In addition, the mucoadhesive material wets, swells and dehydrates (dehydration theory) the mucus, due to the difference of osmotic pressure. The difference in concentration gradient draws the water into the formulation until the osmotic balance is reached. This formulation and mucus are mixed and, therefore, increase the contact time of the

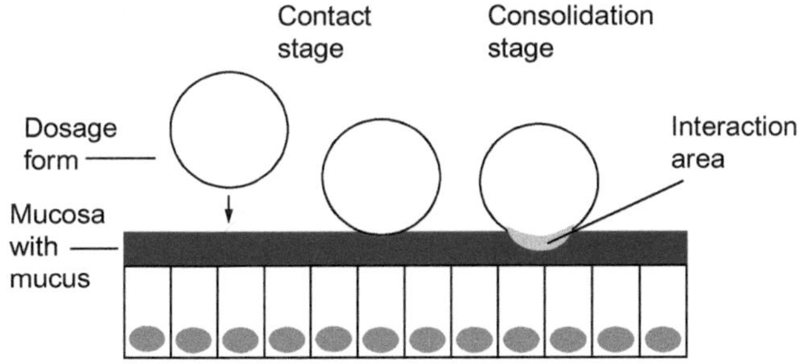

Figure 6.43 The two stages in mucoadhesion.
Reprinted from Smart (2005), with permission from Elsevier.

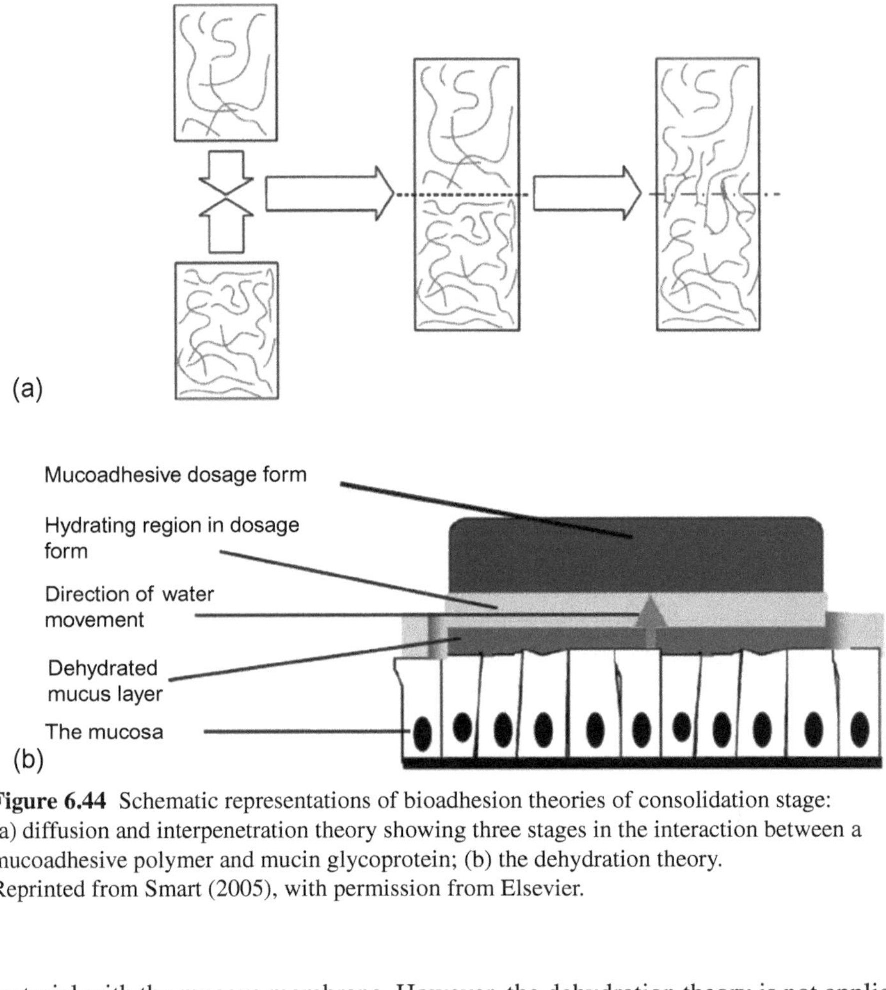

Figure 6.44 Schematic representations of bioadhesion theories of consolidation stage: (a) diffusion and interpenetration theory showing three stages in the interaction between a mucoadhesive polymer and mucin glycoprotein; (b) the dehydration theory.
Reprinted from Smart (2005), with permission from Elsevier.

material with the mucous membrane. However, the dehydration theory is not applicable for highly hydrated systems.

During the process of residence time, the system will have equilibrium between the adhesive forces and the forces of removal. Adhesive failure normally occurs at the weakest component of the joint. When the bioadhesive system is weaker, the failure will be in the mucoadhesive–mucus interface. On the other hand, for stronger bioadhesive materials, this would initially be the mucus layer, but later may be the hydrating mucoadhesive material (Smart, 2005). Therefore, the durability and strength of the adhesion depend on the cohesive nature of the weakest region.

Many mucoadhesive polymers over-hydrate when in an aqueous environment and are readily removed. Thus, it is possible to use cross-linking and introducing hydrophobic entities to control the rate and extent of hydration to render prolonged adhesion. However, every time, the formulations will be removed by mucus or cell turnover.

6.12.2 Bioadhesive materials

The intrinsic properties of the formulation and the environment in which it is applied influence bioadhesive performance (Lee, Park, & Robinson, 2000). Thus, intrinsic factors that affect the bioadhesive performance of a material (normally polymer) are molecular weight, concentration and chain flexibility.

For linear polymers, mucoadhesion increases with molecular weight, but the same relationship does not hold for non-linear polymers. It has been shown that more concentrated mucoadhesive dispersions are retained on the mucous membrane for longer periods, as in the case of systems formed by *in situ* gelification (Lee et al., 2000). After application, such systems spread easily, because they present rheological properties of a liquid, but gelify as they come into contact with the absorption site, thus preventing their rapid removal. Chain flexibility is critical to consolidate the interpenetration between formulation and mucus (Smart, 2005; Carvalho et al., 2010).

Environmental factors affecting the bioadhesive performance of materials include initial contact time, swelling, pH, and physiological variations. Ionizable groups in polymers as well as the formation of charges on the mucus surface are dependent on pH. The extent of chain interpenetration is a result of contact time between the mucus layer and the bioadhesive material. Suitable hydration of the material can lead to good bioadhesive performance. The thickness of the mucus layer can vary from different sites, and other physiological variations can also occur with diseases (Lee et al., 2000; Carvalho et al., 2010).

The first studies about the use of bioadhesive materials were conducted with the purpose of improving the therapy of stomatitis by using adhesive tablets, and increasing the systemic bioavailability of insulin in the form of bioadhesive powder after nasal administration in dogs (Nagai, Nishimoto, Nambu, Suzuki, & Sekine, 1984; Nagai, 1985).

Nowadays, bioadhesive materials are used to increase the residence time of formulation at the site of administration, promote absorption, protect the active agent and control the drug release.

Polymers available on the market (e.g., polyacrylic acids) were the first studied as bioadhesive materials.

The first generation of mucoadhesive materials comprised natural or synthetic hydrophilic molecules containing numerous organic functions that generate hydrogen bonds such as carboxyl, hydroxyl and amino groups, which do not adhere specifically to several surfaces. They were used in dentistry as denture fixers, and the most known examples are carbomers, chitosans, alginates and cellulose derivatives (Bruschi & de Freitas, 2005).

These materials are normally added to solid formulations (e.g., tablets, transdermal adhesives, and micro- and nanoparticles) and into semi-solid formulations (e.g., gels, ointments, pastes, and suppositories) (Smart, 2005). They can be classified into three categories: cationic, anionic and non-ionic.

Cationic molecules have the important property of interaction with the mucus surface, because it is negatively charged at physiological pH. For example, the mucoadhesion of chitosan, a cationic polymer, occurs due to the electrostatic interactions of its amino groups with the sialic groups of mucin in the mucus layer. On the other hand, carbomers (synthetic polymers derived from polyacrylic acid) are negatively charged but are also mucoadhesive. This is explained by the hydrophobic interactions, hydrogen and van der Waals bonds, which are controlled by pH and ionic composition (Woodley, 2001).

Carbomers have been extensively studied as semi-solid mucoadhesive systems (Bruschi & de Freitas, 2005; Bruschi et al., 2007; Jones et al., 2009). They have flexible chains and non-abrasive characteristics when in the partially hydrated state, which decrease the tissue damage caused by friction when they come into contact.

There exist other anionic polymers that are used to prepare bioadhesive systems, like cellulose derivatives (e.g., carboxymethylcellulose) and alginates.

Hydroxypropylmethylcellulose, HEC and methyl cellulose are examples of non-ionic polymers used as bioadhesive materials. They present a weaker mucoadhesion force compared to anionic polymers.

Lyotropics, precursor systems of the crystalline phase, have been suggested as bio-adhesives. When in contact with the water from the environment, the system absorbs water and changes its structure, forming mesophases that increase the residence time of formulation at the site of administration and control the drug release. These systems will be discussed in the proceeding sections.

The second generation of bioadhesive materials is composed of material with the ability to incorporate both hydrophilic and lipophilic active agents, show mucoadhesive properties in its solid and liquid forms, inhibit local enzymes or promote absorption, be specific for a particular cellular area or site, stimulate endocytosis and have a broad safety range (Lee et al., 2000).

They are considered specific bioadhesives because they adhere to specific chemical structures on the cell or mucus surface. Lectins, fimbrial proteins, invasins, antibodies, and those obtained by the addition of thiol groups to known molecules are examples (Bruschi & de Freitas, 2005; Carvalho et al., 2010).

For example, lectins are immunogenic vegetal glycoproteins that specifically recognize sugar molecules. They have the ability to bind (non-covalently) to glycosylated components of the cellular membrane but not of the mucus, and adhesion can therefore be called cytoadhesion. Through the transmission of a cellular signal, this specific bond can result not only in bioadhesion, but also in cellular internalization by different lysosomal and non-lysosomal mechanisms (Lehr, 2000). The most commonly found lectins are those isolated from *Abrus precatroius*, *Agaricus bisporus*, *Anguilla anguilla*, *Arachis hypogaea*, *Pandeiraea simplicifolia* and *Bauhinia purpurea* (Chowdary & Rao, 2004).

It is possible to isolate proteins from the membrane of *Yersinia pseudotuberculosis* that stimulate phagocytosis at the cellular membrane through linkage with integrin receptors. These proteins are named bacterial invasins and are very useful as bioadhesive materials of the second generation.

Antibodies are structures with high specificity and, therefore, can be used as polymeric ligands in the development of site-specific mucoadhesives. This strategy can be useful, for instance, in drugs targeting tumor tissues.

Another type of protein used as bioadhesive material is the group of bacterial fimbrial proteins. They are able to bond to the epithelial surface of erythrocytes, and this property is useful as an efficient mechanism of improving adhesion of mucoadhesive agents used in release systems (Chowdary & Rao, 2004).

The presence of conjugated sulfidryl groups increases the mucoadhesive properties of different polymers. This fact is due to the formation of disulfide bridges with cystein domains of glycoproteins of the mucus. Additionally, these products promote mucus permeation by a mechanism of glutathione regeneration.

The polymers containing these groups are named "thiolated polymers" and normally also display antiprotease activity due to their binding ability with divalent cations, such as zinc and magnesium, which are cofactors for many proteases.

Nowadays, more intelligent mucoadhesive materials are being developed, enabling the acquisition of a unique carrying characteristic for many active agents. The design of these new materials needs to consider the necessity of adhering to any mucous membrane (e.g., vaginal, nasal, buccal, ocular, respiratory, urinary, or gastrointestinal).

6.12.3 Applications

As previously discussed, bioadhesive systems can be designed to be administered by different routes. These bioadhesive systems applied to mucous membranes are frequently defined as mucoadhesive. However, it is possible to develop a bio(muco)adhesive system in different dosage forms, because the properties of adhesion largely depend on the features of the material used in its preparation (Carvalho et al., 2010).

In this context, several conventional drug delivery systems already in use can become bioadhesive after redesign by including bioadhesive substances in their formulation. The strategies previously discussed as well as those that will be discussed can be used together with bioadhesion to improve the characteristics of the systems, like their control of the drug release, and therefore improve the therapeutics.

The main application of the bioadhesion strategy is to modify the drug delivery to control the release of the active within a defined period of time. Thus, in recent years, bioadhesive dosage forms include adhesive solid drug delivery systems (tablets and patches), adhesive semi-solid drug delivery systems and adhesive liquid dispersions (Bruschi & de Freitas, 2005). For example, it is possible to cite solid micro- and nanoparticulate systems, microemulsions, colloidal dispersions of bioadhesive polymers, semi-solid systems, liquid crystalline mesophases, and hydrogels, which can increase the contact time between the preparation and the mucous membrane after they undergo *in situ* gelation (Bruschi et al., 2007, 2008; Bruschi & de Freitas, 2005; Carvalho et al., 2010).

6.12.4 Future and perspectives

Research for the development of new devices and more intelligent materials is a constant in the pharmaceutical sciences. In this context, and considering the great influx of new molecules stemming from drug research, bioadhesive systems play an increasing role in the development of new pharmaceuticals.

The use of bioadhesive materials constitutes a very important strategy to modify the drug delivery, improving the performance of the therapeutic system. The mechanism by which a bioadhesive bond is formed will depend on the nature of the biological membrane and bioadhesive material, type of formulation, attachment process, and subsequent environment of the bond.

Therefore, it is very important to understand the mechanism of bioadhesion and the characteristics of each component of the formulation as well as of the final preparation. For this, various physicochemical and biological tests have been developed and are easily found in the literature. Moreover, with the advance of biotechnology, new materials and possibilities have become a reality.

Oral mucosa bioadhesive systems have been developed to be used in the treatment for aphthae, ulcerations, periodontitis, gingivitis, dental caries, and systemic absorption with the possibility of prolonging residence time, controlling the drug release and improving bioavailability (Bruschi & de Freitas, 2005; Bruschi et al., 2007; 2008; Carvalho et al., 2010).

Drugs with short biological half-lives or stables, requiring a sustained effect and/or exhibiting sensitivity to enzymatic degradation in the intestinal tract, may be successfully delivered via oral bioadhesive delivery systems. The *peroral* and rectal routes are being highly explored for the administration of mucoadhesive systems.

Today, research on the bioadhesive approach as a strategy for modifying drug delivery continues at a rapid pace, aiming for the successful treatment of many diseases.

Optimizing local or systemic treatment of disease via transmucosal drug delivery from the oral, nasal, vaginal and ocular cavities continues to be investigated, using a variety of dosage forms containing novel bioadhesive polymers.

A newly developed bioadhesive coating for nanoparticles has the ability to enhance their intestinal absorption into the bloodstream, offering potential for protein-based medicines such as insulin to be taken orally rather than by injection (Reineke et al., 2013). Bioerodible non-bioadhesive nanospheres were coated with bioadhesive material, poly(butadiene-maleic anhydride-co-L-DOPA) (PBMAD), and the structures showed an enhancement of particle uptake in the small intestine more than 10 times. The surface chemistry plays a significant role in their localization within the body, and the bioadhesive properties enabled the differential distribution of the nanospheres to various tissues. Therefore, the use of bioadhesive polymers could enhance nano- and microparticle uptake from the small intestine for oral drug delivery.

The addition of substances of sensitization and recognition continue to be used for the design of materials with more intelligent mechanisms of mucoadhesion. By binding functional groups within polymer chains, hydrogels can be made more sensitive to surrounding environmental conditions like temperature, moisture, pH, electrical fields, and ionic forces (Bruschi & de Freitas, 2005; Bruschi et al., 2007; 2008; Carvalho et al., 2010).

6.13 Liquid crystalline systems

Liquid crystal systems are formed by amphiphilic molecules that are characterized to possess a polar end that is attracted to water, and a non-polar end that is attracted to hydrocarbons. They are characterized essentially by the presence of the orientation order on the interfaces. In solution, the polar end of amphiphilic molecules is positioned in polar solvent, and the non-polar portion is dissolved in non-polar solvent.

Therefore, they constitute distinct condensed phase materials possessing physical characteristics of solids and melts, with a partial order/disorder of the atomic species. As the concentration of the molecules in the solution increases, they take on different arrangements or phases. For this reason, they are also called mesophases, with the Greek prefix meaning intermediary.

The mesophases originate from the increasing concentration of amphiphilic molecules in the dispersion. At low concentrations, the surfactant is dissolved in the solvent like any other particles of solute distributed randomly throughout the water. When the concentration of amphiphilic molecules reaches the critical micelle concentration, the molecules arrange themselves in micelles under the form of hollow spheres, rods or discs (Figure 6.45).

Besides having varied forms, micelles can also display varied sizes, but the smallest ones have a diameter about twice the length of a hydrocarbon chain with all trans bonds. With the constant addition of amphiphilic molecules to the dispersion, its concentration gradually increases and the micelles become larger and have the ability to dissolve non-polar substances. When the micelles reach a large enough size, the dispersion is no longer a solution, but a coarse dispersion named an emulsion. As previously discussed, depending on the type and amount of amphiphilic molecules added, it is possible to prepare microemulsions that are characterized to possess swollen micelles not large enough to interfere with light, but that are still extremely stable and exist in equilibrium.

6.13.1 Classification

Materials that form liquid crystals come in two basic classifications: lyotropic and thermotropic. Thermotropic liquid crystals depend on temperature, while lyotropic liquid crystals depend mainly on the solvent.

Thermotropic liquid crystals were discovered after the lyotropics. When certain substances (e.g., esters of cholesterol) are heated, they can form thermotropic liquid crystals. This cloudy intermediate phase presents areas with crystal-like molecular structure within the dispersion.

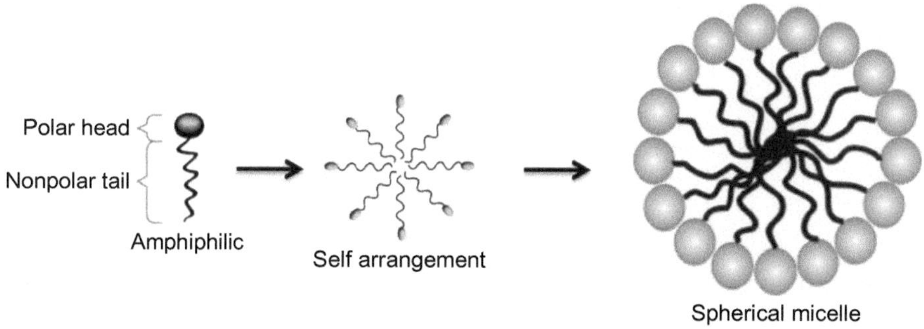

Figure 6.45 Schematic representation of the formation of a spherical micelle.

The surfactants that show this property in water under heating present elongated, flat and rigid structure along their axes. Normally, they are of a variety of chemical types such as azo compounds, azoxy compounds or esters, and the presence of easily polarizable groups often enhances liquid crystal formation (Florence & Attwood, 2006).

There are three principal types of thermotropic liquid crystals: smectic (soap-like), nematic (thread-like), and cholesteric (Figure 6.46).

When the groups of molecules orientate spontaneously with their long axes parallel, but they are not ordered into layers, the state is referred to as nematic liquid crystalline. In the smectic liquid crystalline state, the molecules of surfactant are more ordered than in the nematic. They are arranged with their long axes parallel and are also arranged into distinct layers. This arrangement results in being two-dimensional ordered, viscous, and not orientated by magnetic fields. The cholesteric state is

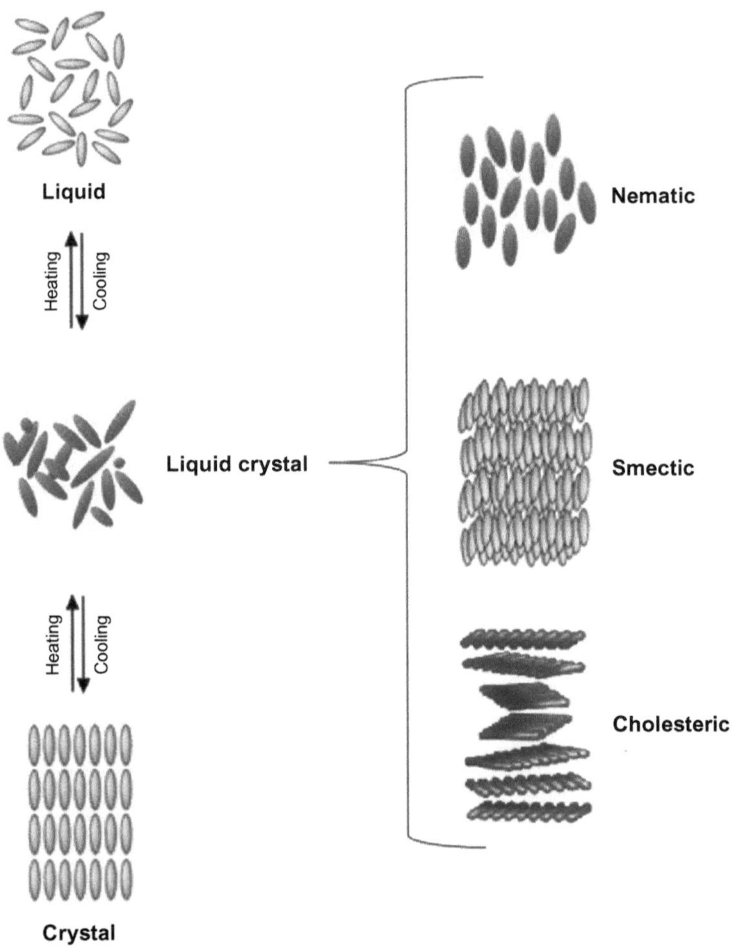

Figure 6.46 Diagrammatic representations of forms of thermotropic liquid crystalline phases.

characterized to display a stack of very thin two-dimensional nematic-like layers in which the elongated molecules lie parallel to each other in the plane of the layer.

Lyotropic mesophases contain at least two components: the organic component (e.g., a surfactant) and its solvent. The organic portion should exhibit some chemical characteristics; otherwise, the solvent would simply solubilize the molecule, yielding molecular solutions of scattered and disordered molecules, and liquid crystals would certainly not form.

The addition of a solvent (e.g., water) will selectively hydrate the hydrophilic portion of the surfactant molecules, preventing hydrophobic regions. Lyotropic mesophases can be considered micelles with ordered molecular arrangement characterized by alternating hydrophobic and hydrophilic regions. As the concentration of surfactant, other liquid crystalline forms may form as lamellar, hexagonal, and cubic.

When a surfactant is dispersed in water, at concentrations close to the critical micellar concentration, the resulting solution is clear and isotropic. In this case, the magnitudes of such physical properties as viscosity and refractive index do not depend on the direction in which these properties are measured.

However, with the increase of the surfactant concentration occurs a transition from the typical spherical micellar structure to a more elongated or rod-like micelle. If the surfactant concentration increases, the orientation and close packing of the elongated micelles into hexagonal arrays may occur.

This new conformation or phase contains ordered arrays and constitutes a liquid crystalline state termed the mesophase or hexagonal phase. The lamellar phase is obtained by increased surfactant concentration. In some surfactant systems, another liquid crystalline state, the cubic phase, occurs between the hexagonal and lamellar phases.

In all systems, the surfactant separates out of the solution and "the liquid crystalline phases that occur on increasing the concentration of surfactant solutions are referred to as lyotropic liquid crystals" (Florence & Attwood, 2006). This mesophases are described in Figure 6.47.

The lamellar phase consists of parallel, planar layers of surfactant bilayers separated by layers of solvent, forming a one-dimensional lattice. In the hexagonal phase, the aggregates consist of long cylinders forming two-dimensional structures. In the

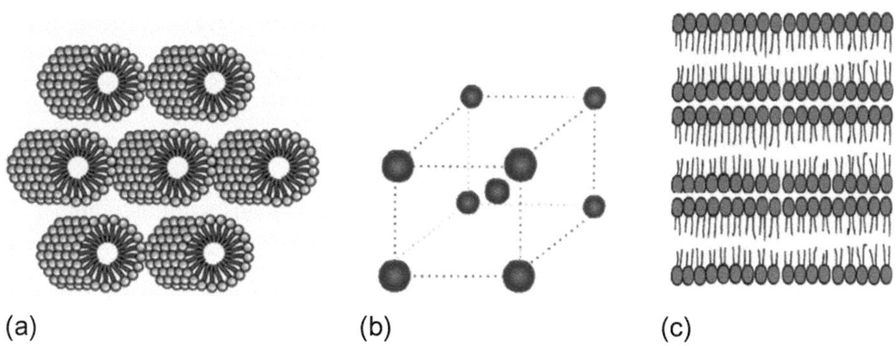

Figure 6.47 Diagrammatic representation of different mesophases of lyotropic liquid crystals: (a) hexagonal mesophase; (b) cubic mesophase; (c) lamellar mesophase.

normal phase, the surfactant molecules aggregate into circular cylindrical micelles, with water filling the volume between the cylinders, while in the reverse phase, cylinders containing the water channel are surrounded by the surfactant polar heads and the oily portion is located around the cylinder.

Lyotropic cubic phases have more complicated structures and are viewed with greater difficulty than the other phases. In almost all of the fluid phases, dimensional cubic symmetry is observed, except in the rhombohedral phase. Reversed topology of the orthorhombic and tetragonal phases can also be detected in some systems.

Hexagonal mesophases are formed by rod-shaped micelles into hexagonal arrays made out of six rods grouped around a central one, for a total of seven. The non-polar groups are isolated inside the micelles. Hexagonal liquid crystals generally exist in solutions that are 40–70% amphiphile.

One way to classify the liquid crystalline phase is to determine the optical isotropy of the mesophase. Under a plane of polarized light, the sample is anisotropic if able to divert the plane of the incident light, and isotropic if not able to deflect light. Lamellar and hexagonal mesophases are anisotropic, while the cubic phases are isotropic.

These mesophases have the characteristics to receive and "accommodate" active agents with different chemical characteristics. The liquid crystals may come apart if too much water or salt is added to the solution, but many varieties can absorb oil by expanding the diameter of the rod-shaped micelle. Therefore, they represent an important strategy to modify the drug delivery of formulations to be administered by different routes.

6.13.2 Uses and applications

There are a high number of applications of lyotropic liquid crystalline formulations. They can be used as soap, working at removing dirt and grease because the non-polar insides of the micelles are capable of dissolving non-polar substances that will not dissolve in water. Soaps also help water dissolve more because the molecules tend to remain at the surface, with the hydrocarbon tail away from the water, thus lowering the surface tension of the water and allowing more material to enter it and be dissolved.

In this sense, many active agents are also more soluble in lyotropic liquid crystals, and these systems can be used as platforms for drug delivery. The liquid crystalline mesophases may be administered by various routes, such as oral, rectal, subcutaneous, topical, transdermal, intramuscular, or intravenous. For oral administration, the drug-containing system must be melted and placed in capsules to be taken and then processed *in situ* in the cubic phase after contact with the gastrointestinal fluids. However, sometimes the system is too viscous for direct parenteral administration (e.g., cubic mesophases), both intramuscular and subcutaneous administration. Thus, it is necessary that the drug delivery system have a low viscosity so that it can be injected and subsequently become a cubic phase.

With regard to topical administration, it is important to state that the outermost layer of skin is primarily a lyotropic liquid crystal made of fatty acids. At least some of the fatty acids must be unsaturated because saturated chains will simply crystallize rather than form a liquid crystal structure. Experiments with crystals made only from saturated fatty acids produced layers that did not prevent water transport across them. This is a symptom that sometimes occurs in people whose diets are deficient in essential fatty acids of the sort found in unsaturated fats. In this context, higher concentrations of hydrocortisone can be topically administered when this drug is incorporated into a liquid crystal system (Savic et al., 2007).

Liquid crystalline phases that deliver small interfering RNAs (siRNAs; potently but reversibly silent genes *in vivo* that interfere with disease-causing or disease-promoting genes) were proposed to be administered topically. The cationic polymer polyethylenimine or the cationic lipid oleylamine was added to monoolein-based systems in different concentrations. The results showed that the studied nanodispersions may represent a promising new non-viral vehicle and can be considered highly advantageous in the treatment of skin disorders. They were effective in optimizing the skin penetration of siRNA and reducing the levels of the model protein GAPDH without causing skin irritation (Vicentini et al., 2013).

Liquid crystalline systems can be used to modify the release of active agents. Lyotropic liquid crystal can coat an active to keep it from being destroyed in the digestive tract. The active can then be taken orally, and after it reaches the proper location in the body, the liquid crystal breaks down and the drug is released.

Ester groups of fatty acids, such as glyceryl monooleate and glyceryl monolinoleate, are able to build liquid crystals that in turn can act as controlled-release systems. These fatty acids build lyotropic liquid crystalline mesophases in the presence of water at body temperature.

The cubic phase favors the controlled release of active agents, because it has a structure made up of tridimensional curved lipid bilayers, separated by congruent water channels. This structure has the appearance of highly viscous transparent gel. Due to this relatively high viscosity, it is difficult to administer on any mucous membrane. Less viscous mesophases such as the lamellar phase can be used to circumvent these problems. In these instances, this phase is considered a precursor of the cubic phase. In the case of lyotropic mesophases, the precursor absorbs water *in situ* and spontaneously builds the cubic phase. This was observed using the surfactant PPG-5-Ceteth-20, isopropyl myristate, and water to deliver propolis for the treatment of periodontal diseases. The rheological and textural characteristics of liquid crystalline precursor systems were determined, revealing characteristics that favored easy insertion into the periodontal pocket and subsequent stable retention therein. The release profile studies showed that the propolis can be released from the systems for a prolonged period of time greater than 7 days (Bruschi et al., 2008).

The inverse bicontinuous cubic phase is of particular interest in drug delivery. It is an isotropic, thermodynamically stable and highly viscous phase containing congruent networks of water channels. Therefore, this liquid crystalline phase can enclose hydrophilic, amphiphilic and active agents (e.g., proteins, peptides, amino acids, and nucleic acids). There are three forms of cubic phase: cubic phase gel, cubic phase precursor, and cubosomes. Cubic phase gels have been commonly used in mucosal, vaginal, periodontal and transdermal drug delivery. However, the stiffness and viscous nature of cubic phase gels limit their potential application as a delivery system. Precursor systems of the cubic phase have been proposed for arterial transcatheter chemoembolization on hepatocellular carcinoma. Cubosomes are discrete, submicron or nanostructured particles formed from fragmentation and steric stabilization of inverse bicontinuous cubic phases of lipids. They possess much larger specific surface area and still remain the inner structures and the sustained release property and mechanism of the cubic phase as a delivery system. Their properties are favorable for using the cubosomes as a universal platform for drug delivery (Pan et al., 2013).

In addition, mucoadhesion for formulations formed by liquid crystals can be explained by their rheological properties. These properties decrease the mucociliar clearance and increase the contact time of the formulation with the mucous membrane (Carvalho et al., 2010).

6.13.3 Future and perspectives

Drug–system interactions play an important role in controlling the release. However, the release rate of incorporated drugs in liquid crystalline mesophase systems depends on the structure and physicochemical properties of the active agent. These properties make possible the use of liquid crystals as drug carrier platforms, which may be able to control the release of the substances incorporated therein. As a result of active agent–solvent interaction, under certain conditions, the solubilized actives can induce changes in the molecular packing of the system. This effect may influence the rate of *in vitro* release.

Hydrogels of poloxamer 407 have the ability to form thermotropic liquid crystals with high viscosity. They are used as controlled drug delivery by rectal and topical routes. They can work as reservoir systems for indomethacin, which is released by diffusion through the water channels contained in the matrix (Miyazaki et al., 1986).

Liquid crystal precursor formulation administered by the nasal route showed itself to be a promising novel tool for the systemic delivery of zidovudine (AZT) and other antiretroviral drugs (Carvalho, Campos, Peccinini, & Gremião, 2013). A low-viscosity precursor of a liquid crystal phase was developed using PPG-5-CETETH-20, oleic acid, and water (55%, 30%, 15% w/w). Using the advantage of lyotropic phase behavior, a mucoadhesive stimuli-sensitive drug delivery system was developed. The formulation could form a mucoadhesive matrix in contact with nasal mucus that promoted nasal absorption of the AZT. The Weibull mathematical model suggested that the AZT is released by Fickian diffusion mechanisms.

6.14 Implants

The elastomers discovery starts the application of biocompatible polymers as a strategy for the development of implantable platforms for drug delivery. These systems have the advantage of better control of delivery during the period of duration of the active agent. Moreover, the dosages are more precise.

During the 1960s, silicon elastomers and other biocompatible polymers were exhaustively investigated as materials to control the drug delivery of hydrophilic molecules. In 1975, the first silicone implants were proposed by Folkman for slow drug release, as discussed in previous chapters. The first investigations showed that the drug delivery rate was controlled by thickness and area of membrane surface, polarity of penetrant, and the characteristics of used material.

Implants are defined as sterile drug delivery systems obtained by compression, melting, molding or sintering, and consisting basically of the active agent and the rate-controlling excipients. These systems are solid dosage forms (about 3.2 mm in diameter and 8 mm long) destined to be implanted or administered by subcutaneous route.

The solid systems can be implanted under the skin, usually of the thigh or abdomen, using a special injector or by surgical incision. They are termed pellets and enable the drug release at a fairly constant rate over several weeks to many months.

6.14.1 Vantages and advantages

The first generation of implants was non-biodegradable systems and offered the advantages of long-lasting drug release and reduced host response. With the evolution of the material sciences, new eroding polymers enable the development of implants with the ability to sustain the drug release, and improved therapeutics of active agents.

Using biodegradable materials, it is possible to develop implants with controlled erosion rate and spontaneous degradation, allowing the modulation of suitable or desired kinetics of drug release.

Besides the solid implants, biodegradable polymers can be used to form injectable implants, or they can be used to encapsulate nano or microparticles, as previously discussed. Particulate systems can be injected through thin needles, and have different behavior and distribution in the site of administration associated with their size and composition.

For example, polymers can be used to prepare viscous or semi-solid systems that can be localized within the eye and used as a slow-release intraocular "implant" after a simple injection. Moreover, responsive polymers can be used and appropriately delivered as needed by localized application of heat or various light lengths (Bruschi et al., 2007; Jones et al., 2009; Bourges et al., 2006).

Therefore, an increasing number of types of implants are being proposed as a strategy to modify drug delivery. Mini and micropumps can be implanted under the skin and, like these, some strategies were also discussed in previous sections of this chapter. Therefore, the main implanted systems discussed in this section are the solid implants.

The device is completely subcutaneous, with no opening in the skin. Therefore, there is little chance of infection or interference with daily activities. Moreover, these systems or devices offer several advantages over conventional oral or parenteral dosage forms:

- They allow site-specific administration of an active agent (e.g., implants used in the treatment of brain tumors or prostate cancer).
- They enable the modified and controlled delivery of an active agent.
- They allow targeted drug delivery.
- They decrease the side effects.
- They increase the bioavailability of active agents.
- They increase patient compliance, because the treatment regimen is associated with few (or just one) administrations.

On the other hand, mini-surgery is needed to implant the system, it is not easy to simply discontinue the therapy, and local reactions in the patient can occur. In this context, some of the critical questions for ongoing research on implants have been concerned with irritation/carcinogenicity, reproducibility of erodibility, dose dumping, duration, and pulses.

6.14.2 Uses and applications

Subcutaneous solid implants enable long-term passive drug release without the need for replenishment. The simplest implant systems are found as thin flexible rods or cylinders, which are effective for the delivery of highly potent active agents such as hormones (e.g., histrelin for the palliative treatment of prostate cancer and uterine fibroids, and early puberty in children; levonorgestrel and etonogestrel for family planning). Actually, implant systems are commercially available for contraception and hormone replacement therapy, and to improve the rate of *in vitro* fertilization. There are other uses of implants for delivery of active agents used in the treatment of schizophrenia, breast cancer, photosensitivity, and Parkinson's disease.

Particularly in ophthalmology, there are a number of promising implant systems proposed for the treatment of macular edema and retinal vein occlusion using corticosteroids, with products in development for the treatment of glaucoma and age-related macular degeneration. Non-biodegradable (e.g., polyvinyl alcohol—EVA and polysulphone capillary fiber) and biodegradable solid implants (e.g., poly lactic acid, poly glycolic acid, poly lactic-co-glycolic acid, polycaprolactones, and polyanhydrides) as well as viscous and injectable polymers (e.g., poly(ortho esters)) have been used for therapeutic ocular implants (Bourges et al., 2006).

The main mechanisms by which the drug delivery rate is controlled are the diffusion-controlled mechanism, solvent-controlled (osmotic) release, erodible materials, and regulated release mechanisms (e.g., externally applied trigger to turn release on/off by electrical or mechanical stimulus).

6.14.3 Materials

Because the implant has the aim to be implanted and to keep releasing the active agent for a period of time and under a controlled rate, the materials used need to be biocompatible, stable, and durable (except in the case of biodegradable drug delivery systems). Moreover, the materials should control the release of the active agent.

Silicones were first used due to their chemical inertness, range of stable mechanical properties, and ability to compound various active agents together in the system. For this type of silicone implant, the drug release rate is generally proportional to the loading of the drug within the silicone (from 5% to 50%, w/w).

Another material used is EVA, a stable material with the additional ability to control the release rate by varying the vinyl acetate content. Polyurethanes and acrylate hydrogels have also been utilized in select applications.

The method of preparation of the solid implants is dependent on their size, materials and the active agent. The main methodologies are based on molding or extrusion, and many factors should be considered in optimizing the material formulation and developing a robust molding or extrusion process: the active agent, materials, mixing or compounding technique, used temperature, agitation and pressure. These variables can interfere in the performance of the implant system.

6.15 Colon-specific delivery systems

The large intestine is not traditionally considered an absorption organ of active agents, with the exception of the terminal portion (rectum), which is utilized for the administration of some dosage forms, such as suppositories and ointments. These traditional dosage forms are destined for local and systemic treatments, allowed by connection between hemorrhoidal veins and the general bloodstream.

The administration of conventional dosage forms by *peroral* route conduces to dissolution of the dosage form in the gastrointestinal fluid and, depending on the physicochemical characteristics, the consequent absorption of the active agent.

The colonic region is susceptible to many diseases, like inflammatory diseases (e.g., ulcerative colitis, Crohn's disease), infectious diseases (e.g., spasmodic colitis; amoebian, bacterian, viral, schemic, and Chagas diseases) and carcinomas. Therefore, when the active agent needs to be delivered only on the colon or an active agent needs to be protected from the gastrointestinal environment, oral administration using conventional dosage forms is a problem.

The ascending, descending and transverse colon have been much studied as a place for drug administration for local and systemic therapy. The colonic delivery of an active agent can be defined as a type of modified delivery, where the targeted site is the colon.

When the drug is delivered into the colonic region rather than the upper gastrointestinal region, this strategy is useful for the treatment of Crohn's disease, ulcerative colitis, carcinomas, and infections. Moreover, some therapeutic agents useful for the treatment of colonic diseases have been overlooked due to their toxicity. However, controlled delivery of the active agent at the target site can allow their use.

Colonic delivery is an attractive strategy to improve the bioavailability of poorly absorbed drug molecules. The presence of numerous bacteria in the colon enables prodrugs and formulations to be degraded, controlling the drug delivery. This approach is of particular interest for active agents that are acting as a substrate for cytochrome P450 3A, because activity of these classes of enzymes is comparatively lower in the colonic portion (Friend, 2004).

6.15.1 Advantages and disadvantages

Colonic drug targeting possesses advantages for active agents that are unstable in the upper gastrointestinal tract environment (e.g., by stomach acid, by pancreatic enzymes).

This strategy provides better patient compliance due to the reduction in dose and dosage frequency. Moreover, it provides a more effective therapy of diseases like colorectal cancer, ulcerative colitis, and Crohn's disease by delivery of drugs to the infected area.

Therefore, colonic-targeted delivery systems can protect the active agent from lower pH from the stomach and in the case of diseases (e.g., bowel disease where the pH of the colon decreases below normal), and start to release the active to a certain extent in the lower intestinal tract.

In addition, some drugs can be absorbed in the colon, such as bromofeniramine, diclofenac, glibenclamide, olsalazin, ibuprofen, isosorbide nitrate, nifedipine, teophiline, heparin, and vasopressin.

Oral administration of peptides for systemic absorption has increased the research on colonic delivery as well. The bioavailability of these drugs is made difficult by the easy hydrolysis they suffer in the upper portion of the gastrointestinal tract. The colonic controlled delivery can decrease this problem because the peptides are delivered in a site of less enzymatic activity. The colonic site possesses 80% less enzymatic activity than the small intestine.

Moreover, chronotherapy or the treatment of some diseases that show variation during the period of a day (e.g., asthma, inflammation, hypertension, arthritis, or cardiac arrhythmias) can be treated using this strategy. Asthma is generally less controlled at night and worsens in the morning. A dose administered when the patient is going to sleep can enable systemic absorption, due to the period of transit, when necessary.

On the other hand, therapeutic systems for colonic delivery are dependent on some factors (Bansal, Malviya, Malaviya, & Sharma, 2014). If the system is designed for localized treatment, then its pathological conditions and disease pattern should be considered. The physiochemical and biopharmaceutical properties of the active agent need to be considered, as well as the release characteristics of the active agent. As previously discussed, the pH gradient throughout the gastrointestinal tract, transit of materials into and through the colon, effect of diet on colonic transit and colonic bacteria are factors that should be considered as well.

6.15.2 Uses and applications

Colonic drug delivery can be used to treat various types of diseases. Therefore, the best candidates for colon targeting are active agents that display less absorption from the upper portion of the gastrointestinal tract.

Colonic-targeted drug delivery is useful for the treatment of inflammatory bowel diseases. In this case, the systems have been developed using pH-dependent delivery, enzyme-dependent delivery (for prodrugs, matrices, and coatings), pulsatile time-dependent delivery, and pressure-dependent delivery approaches (Bansal et al., 2014).

Nowadays, some of these approaches are being used together. It is possible to combine pH and microbial dependence with pulsatile time delivery. For example, a tablet core of lactulose coated with an acid-soluble polymer (Eudragit E) is recoated with an enteric polymer (Eudragit L). This last coating protects the tablet from the upper part of the gastrointestinal tract and dissolves in the small intestine. When the formulation comes into contact with the colonic fluid, polysaccharide (lactulose) is enzymatically degraded by the colonic microfloras into organic acid. The pH around the formulation decreases and the release of the active agent takes place due to the dissolution of the acid-soluble coating (Masataka et al., 2004).

Osmotically controlled drug delivery systems can be developed for the colon. In this case, a semi-permeable membrane with an osmotic agent and the active agent are used to deliver the drug to the colonic region. An orifice is drilled through the membrane next to the drug layer. This membrane is then coated with an enteric-coated polymer to prevent the drug from being released in the upper gastrointestinal tract. The enteric polymer dissolves due to the higher pH of the small intestine. The water moves toward the core, increasing the volume within the osmotic compartment, forcing the active agent out of the devices through the delivery, as previously discussed for osmotic pumps.

6.15.3 Materials

Many synthetic polymers used for colonic drug delivery have already been discussed. In addition, natural polymers have been applied for the targeting of active agents into the colon. Disaccharides (e.g., lactose and maltose), oligosaccharides (e.g., cellobiose, cyclodextrins, lactulose, raffinose, and stachyose), and polysaccharides (e.g., alginates, pectin, dextran, unulin, dellulose, and arabinogalactan) are examples of these natural polymers.

6.15.4 Future and perspectives

Colonic-targeted drug delivery is extremely dependent on the concentration of colonic microflora, which can be altered in disease conditions. The understanding about the behavior of the polymers that are degraded by colonic bacteria should be studied in the disease state for future prospects.

However, considering all factors previously discussed, colonic delivery constitutes an important strategy for the delivery of acid-labile active agents as well as proteins and peptides.

References

Abu-Diak, O. A., Andrews, G. P., & Jones, D. S. (2012). Hydrophobic polymers of pharmaceutical significance. In J. Siepmann, R. A. Siegel, & M. J. Rathbone (Eds.), *Fundamentals and applications of controlled release drug delivery* (pp. 47–73). Springer: New York.

Allen, T. M., & Cullis, T. R. (2013). Liposomal drug delivery systems: From concept to clinical applications. *Advanced Drug Delivery Reviews*, *65*, 36–48.

Allen, L. V., Jr., Popovich, N. G., & Ansel, H. C. (2011). Ansel´s pharmaceutical dosage forms and drug delivery. Baltimore-Philadelphia: Lippincott Williams & Wilkins, 1–710.

Amkraut, A., Eckenhoff, J. B., & Nichols, K. (1990). Osmotic delivery of peptides and macromolecules. *Advanced Drug Delivery Reviews*, *4*, 255–276.

Atwood, D. (2001). Disperse systems. In M. E. Aulton (Ed.), *Pharmaceutics: The science of dosage form design* (2nd ed., pp. 91–93). London: Churchill Livingstone.

Azagury, A., Khoury, L., Enden, G., & Kost, J. (2014). Ultrasound mediated transdermal drug delivery. *Advanced Drug Delivery Reviews*, *72*, 127–143.

Bangham, A. D., Standish, M. M., & Watkins, J. C. (1965). Diffusion of univalent ions across the lamellae of swollen phospholipids. *Journal of Molecular Biology*, *13*, 238–252.

Banker, G. S., & Anderson, N. S. (2001). Tablets. In L. Lachman, H. A. Lieberman, & J. L. Kanig (Eds.), *The theory and practice of industrial pharmacy* (pp. 509–597). Lisbon: Fundation Calouste Gulbekian.

Bansal, V., Malviya, R., Malaviya, T., & Sharma, P. K. (2014). Novel prospective in colon specific drug delivery system. *Polimery w Medycynie*, *44*(2), 109–118.

Bardonnet, P. L., Faivre, V., Pugh, W. J., Piffaretti, J. C., & Falson, F. (2006). Gastroretentive dosage forms: Overview and special case of *Helicobacter pylori*. *Journal of Controlled Release*, *111*, 1–18.

Benita, S. (2006). *Microencapsulation—Methods and industrial applications*, (pp. 1–741). Boca Raton: CRC Press.

Bourges, J. L., Bloquel, C., Thomas, A., Froussart, F., Bochot, A., Azan, F., et al. (2006). Intraocular implants for extended drug delivery: Therapeutic applications. *Advanced Drug Delivery Reviews*, *58*, 1182–1202.

Brandau, T. (2002). Preparation of monodisperse controlled release microcapsules. *International Journal of Pharmaceutics*, *242*, 179–184.

Bruschi, M. L., & de Freitas, O. (2005). Oral bioadhesive drug delivery systems. *Drug Development and Industrial Pharmacy*, *31*(3), 293–310.

Bruschi, M. L., Jones, D. S., Panzeri, H., Gremião, M. P. D., de Freitas, O., & Lara, E. H. G. (2007). Semisolid systems containing propolis for the treatment of periodontal disease: In vitro release kinetics, syringeability, rheological, textural, and mucoadhesive properties. *Journal of Pharmaceutical Sciences*, *96*(8), 2074–2089.

Bruschi, M. L., Jones, D. S., Panzeri, H., Gremião, M. P. D., Freitas, O., & Lara, E. H. G. (2008). Development and characterization of precursor of liquid crystalline phase with propolis-containing microparticles for use in the treatment of periodontal disease. *Drug Development and Industrial Pharmacy*, *34*(3), 267–278.

Buri, P. (1987). Définition et classification des systèmes matriciels. *STP Pharma*, *3*(3), 193–199.

Carvalho, F. C., Bruschi, M. L., Evangelista, R. C., & Gremião, M. P. D. (2010). Mucoadhesive drug delivery systems. *Brazilian Journal of Pharmaceutical Sciences*, *46*(1), 1–18.

Carvalho, F. C., Campos, M. L., Peccinini, R. G., & Gremião, M. P. (2013). Nasal administration of liquid crystal precursor mucoadhesive vehicle as an alternative antiretroviral therapy. *European Journal of Pharmaceutics and Biopharmaceutics*, *84*(1), 219–227.

Chang, R. K. (1992). Formulation approaches for sustained release oral suspensions. *Pharmaceutical Technology*, *16*, 134–136.

Chowdary, C. P. R., & Rao, Y. S. (2004). Mucoadhesive microspheres for controlled drug delivery. *Biological & Pharmaceutical Bulletin*, *27*(11), 1717–1724.

Chung, J. E., Yokoyama, M., Yamato, M., Aoyagi, T., Sakurai, Y., & Okano, T. (1999). Thermoresponsive drug delivery from polymeric micelles constructed using block copolymers of poly(N-isopropylacrylamide) and poly(butylmethacrylate). *Journal of Controlled Release*, *62*, 115–127.

Collett, J., & Moreton, C. (2001). Modified-release peroral dosage form. In M. E. Aulton (Ed.), *Pharmaceutics: The science of dosage form design* (2nd ed., pp. 289–306). London: Churchill Livingstone.

Danielsson, I., & Lindamn, B. (1981). The definition of microemulsion. *Colloids Surfaces*, *3*, 391–392.

Dash, T. R., & Verma, P. (2013). Matrix tablets: An approach towards oral extended release drug delivery. *International Journal of Pharma Research & Review*, *2*(2), 12–24.

Falkiewicz, M. J. (1996). Theory of suspensions. In H. A. Lieberman, M. M. Rieger, & G. S. Banker (Eds.), *Pharmaceutical dosage forms: Disperse systems* (2nd ed., Vol. 1, pp. 17–50). New York: Marcel Dekker.

Fara, J. W., & Ray, N. (1988). Osmotic pumps. In P. Tyle (Ed.), *Drug delivery devices* (pp. 137–175). New York: Marcel Dekker.

Ferrero, C., Massuelle, D., & Doelker, E. (2010). Towards elucidation of the drug release mechanism from compressed hydrophilic matrices made of cellulose ethers. II. Evaluation of a possible swelling-controlled drug release mechanism using dimensionless analysis. *Journal of Controlled Release*, *141*, 223–233.

Ferrero, C., Massuelle, D., Jeannerat, D., & Doelker, E. (2008). Towards elucidation of the drug release mechanism from compressed hydrophilic matrices made of cellulose ethers. I. Pulse-field-gradient spin-echo NMR study of sodium salicylate diffusivity in swollen hydrogels with respect to polymer matrix physical structure. *Journal of Controlled Release*, *128*, 71–79.

Ferrero, C., Massuelle, D., Jeannerat, D., & Doelker, E. (2013). Towards elucidation of the drug release mechanism from compressed hydrophilic matrices made of cellulose ethers. III. Critical use of thermodynamic parameters of activation for modeling the water penetration and drug release processes. *Journal of Controlled Release*, *170*, 175–182.

Florence, A. T., & Attwood, D. (2006). *Physicochemical principles of pharmacy* (4th ed.). London: Pharmaceutical Press, 1–492.

Friend, R. D. (2004). New oral delivery systems for treatment of inflammatory bowel disease. *Advanced Drug Delivery Reviews*, *57*, 247–265.

Gao, P., Rush, B. D., Pfund, W. P., Huang, T., Bauer, J. M., Morozowich, W., et al. (2003). Development of a supersaturable SEDDS (S-SEDDS) formulation of paclitaxel with improved oral bioavailability. *Journal of Pharmaceutical Sciences*, *92*, 2386–2398.

Ghosh, T., & Ghosh, A. (2011). Drug delivery through osmotic systems—an overview. *Journal of Applied Pharmaceutical Science*, *1*(2), 38–49.

Gustafsson, J., Ljusberg-Wahren, H., Almgren, M., & Larsson, K. (1996). Cubic lipid–water.

Hägerström, H., Edsman, K., & Strømme, M. (2003). Low- frequency dielectric spectroscopy as a tool for studying the compatibility between pharmaceutical gels and mucus tissue. *Journal of Pharmaceutical Sciences*, *92*(9), 1869–1881.

Higuchi, T. (1961). Rate of release of medicaments from ointments bases containing drugs in suspension. *Journal of Pharmaceutical Sciences*, *50*(10), 874–875.

Higuchi, W. I. (1962). Analysis of data on the medicament release from ointments. *Journal of Pharmaceutical Sciences*, *51*(8), 802–804.

Higuchi, T. (1963). Mechanism of sustained-action medication. *Journal of Pharmaceutical Sciences*, *52*(12), 1145–1149.

Higuchi, T. & Leeper, H.M. (1973). ALZA Corp., assignee. Osmotic dispenser. US Patent 3,732,865.

Hoar, T. P., & Schulman, J. H. (1943). Transparent water in oil dispersions: Oleopathic hydromicelle. *Nature*, *152*, 102.

Ichikawa, M., Watanabe, S., & Miyake, Y. (1991). A new multiple-unit oral floating dosage system. I: Preparation and in vitro evaluation of floating and sustained-release characteristics. *Journal of Pharmaceutical Sciences*, *80*(11), 1062–1066.

International Union of Pure and Applied Chemistry. (1972). *Manual of colloid science*. London: Butterworth.

Jain, N. K. (1997). Controlled and novel drug delivery. New Delhi: CBS Publisher, 236–237.

Jamekhorshid, A., Sadrameli, S. M., & Farid, M. (2014). A review of microencapsulation methods of phase change materials (PCMs) as a thermal energy storage (TES) medium. *Renewable & Sustainable Energy Reviews*, *31*, 531–542.

Jones, D. S. (2008). *Pharmaceutics—Dosage form and design*. London: Pharmaceutical Press, pp. 25–42.

Jones, D. S., Bruschi, M. L., Freitas, O., Gremião, M. P. D., Lara, E. H. G., & Andrews, G. P. (2009). Rheological, mechanical and mucoadhesive properties of thermoresponsive, bioadhesive binary mixtures composed of poloxamer 407 and carbopol 934P designed as platforms for implantable drug delivery systems for use in the oral cavity. *International Journal of Pharmaceutics*, *372*, 49–58.

Kim, Y.-C., Park, J.-H., & Prausnitz, M. R. (2012). Microneedles for drug and vaccine delivery. *Advanced Drug Delivery Reviews*, *64*(14), 1547–1568.

Kreilgaard, M. (2002). Influence of microemulsions on cutaneous drug delivery. *Advanced Drug Delivery Reviews*, *54*(S1), S77–S98.

Kurkov, S. V., & Loftsson, T. (2013). Cyclodextrins. *International Journal of Pharmaceutics*, *453*, 167–180.

Langer, R. (2004). Transdermal drug delivery: Past progress, current status, and future prospects. *Advanced Drug Delivery Reviews*, *56*, 557–558.

Lawrence, M. J., & Rees, G. D. (2012). Microemulsion-based media as novel drug delivery systems. *Advanced Drug Delivery Reviews*, *64*(S), 175–193.

Lee, J. W., Park, J. H., & Robinson, J. R. (2000). Bioadhesive-based dosage forms: The next generation. *Journal of Pharmaceutical Sciences*, *89*(7), 850–866.

Lehr, C. M. (2000). Lectin-mediated drug delivery: The second generation of bioadhesives. *Journal of Controlled Release*, *65*(1), 19–29.

Loftsson, T., & Duchêne, D. (2007). Cyclodextrins and their pharmaceutical applications. *International Journal of Pharmaceutics*, *329*, 1–11.

Lovelyn, C., & Attama, A. A. (2011). Current state of nanoemulsions in drug delivery. *Journal of Biomaterials and Nanobiotechnology*, *2*, 626–639.

Masataka, K., Watanabe, S., Takemura, S., Sako, K., Sawada, T., Masuda, Y., et al. (2004). Scintigraphic evaluation of a novel colon-targeted delivery system (CODESTM) in healthy volunteers. *Journal of Pharmaceutical Sciences*, *93*(5), 1287–1299.

Mason, T. G., Wilking, J. N., Meleson, K., Chang, C. B., & Graves, S. M. (2006). Nanoemulsions: Formation, structure, and physical properties. *Journal of Physics. Condensed Matter*, *18*, R635–R666.

McClements, D. J. (2012). Nanoemulsions versus microemulsions: Terminology, differences, and similarities. *Soft Matter*, *8*, 1719–1729.

Miyazaki, S., Yokouchi, D., Nakamura, T., Hashigushi, N., Hou, W. M., & Takada, M. (1986). Pluronic F-127 gels as a novel vehicle for rectal administration of indomethacin. *Chemical & Pharmaceutical Bulletin*, *34*, 1801–1808.

Moghassemi, S., & Hadjizadeh, A. (2014). Nano-niosomes as nanoscale drug delivery systems: An illustrated review. *Journal of Controlled Release*, *185*, 22–36.

Muller-Goymann, C. C., & Hamann, H.-J. (1993). Sustained release from reverse micellar solutions by phase transformations into lamellar liquid crystals. *Journal of Controlled Release*, *23*, 165–174.

Munday, D. C., & Cox, P. J. (2000). Compressed xanthan and karaya gum matrices: Hydration, erosion and drug release mechanisms. *International Journal of Pharmaceutics*, *203*, 179–192.

Nagai, T. (1985). Adhesive topical drug delivery system. *Journal of Controlled Release*, *2*, 121–134.

Nagai, T., Nishimoto, Y., Nambu, N., Suzuki, Y., & Sekine, K. (1984). Powder dosage form of insulin for nasal administration. *Journal of Controlled Release*, *1*(1), 15–22.

Nash, R. A. (1988). Pharmaceutical suspensions. In H. A. Lieberman, M. M. Rieger, & G. S. Banker (Eds.), *Pharmaceutical dosage forms: Disperse systems*: Vol. 1. (pp. 151). New York: Marcel Dekker.

Ochoa, M., Mousoulis, C., & Ziaie, B. (2012). Polymeric microdevices for transdermal and subcutaneous drug delivery. *Advanced Drug Delivery Reviews*, *64*(14), 1603–1616.

Olsen, K. M., & Hitzeman, M. L. (2009). Dexlansoprazole MR in the management of gastroesophageal reflux disease. *Clinical Medicine: Therapeutics*, *1*, 1641–1652.

Pan, X., Han, K., Peng, X., Yang, Z., Qin, L., Zhu, C., et al. (2013). Nanostructed cubosomes as advanced drug delivery system. *Current Pharmaceutical Design*, *19*, 6290–6297.

Patil, N. S., Dordick, J. S., & Rethwisch, D. G. (1996). Macroporous poly(sucrose acrylate) hydrogel for controlled release of macromolecules. *Biomaterials*, *17*, 2343–2350.

Peppas, N. A., & Buri, P. A. (1985). Surface, interfacial and molecular aspects of polymer bioadhesion on soft tissue. *Journal of Controlled Release*, *2*, 257–275.

Pereira, R. R. A., Godoy, J. S. R., Svidzinski, T. I. E., & Bruschi, M. L. (2013). preparation and characterization of mucoadhesive thermoresponsive systems containing propolis for the treatment of vulvovaginal candidiasis. *Journal of Pharmaceutical Sciences*, *102*(4), 1222–1234.

Prince, L. M. (1967). A Theory of aqueous emulsions-I. Negative interfacial tension at the oil/water interface. *Journal of Colloid and Interface Science*, *23*, 165–173.

Reineke, J., Cho, D. Y., Dingle, Y. L., Cheifetz, P., Laulicht, B., Lavin, D., et al. (2013). Can bioadhesive nanoparticles allow for more effective particle uptake from the small intestine? *Journal of Controlled Release*, *170*, 477–484.

Rogers, T. L., Hu, J., Yu, Z., Johnston, K. P., Williams, R. O., III (2002). A novel particle engineering technology: Spray-freezing into liquid. *International Journal of Pharmaceutics*, *242*(1-2), 93–100.

Rose, S., & Nelson, J. F. (1955). A continuous long-term injector. *The Australian Journal of Experimental Biology and Medical Science*, *33*(4), 415–419.

Rowe, C. R., Sheskey, P. J., & Quinn, M. E. (2009). *Handbook of pharmaceutical excipients* (6th ed.). London: Pharmaceutical Press.

Santus, G., & Baker, R. W. (1995). Osmotic drug delivery: A review of the patent literature. *Journal of Controlled Release*, *35*, 1–21.

Savic, S., SaviC, M., Tamburic, S., Vuleta, G., Vesic, S., & Müller-Goymanne, C. (2007). An alkylpolyglucoside surfactant as a prospective pharmaceutical excipient for topical formulations: The influence of oil polarity on the colloidal structure and hydrocortisone in vitro/in vivo permeation. *European Journal of Pharmaceutical Sciences*, *30*, 441–450.

Scherlund, M., Malmsten, M., Holmqvist, P., & Brodin, A. (2000). Thermosetting microemulsions as delivery systems for periodontal anesthesia. *International Journal of Pharmaceutics*, *194*, 103–116.

Schulman, J. H., Stoeckenius, W., & Prince, L. M. (1959). Mechanism of formation and structure of microemulsions by electron microscopy. *Journal of Physical Chemistry*, *63*, 1677–1680.

Siepmann, J., & Peppas, N. A. (2001). Modeling of drug release from delivery systems based on hydroxypropyl methylcellulose (HPMC). *Advanced Drug Delivery Reviews*, *48*(2–3), 139–157.

Smart, J. D. (2005). The basics and underlying mechanisms of mucoadhesion. *Advanced Drug Delivery Reviews*, *57*(11), 1556–1568.

Solans, C., & Solé, I. (2012). Nano-emulsions: Formation by low-energy methods. *Current Opinion in Colloid & Interface Science*, *17*, 246–254.

Stegemann, S. (2011). Capsules as a delivery system for modified-release products. In C. G. Wilson, & P. J. Crowley (Eds.), *Controlled release in oral drug delivery* (pp. 277–298). London: Springer.

Stella, V. J., & He, Q. (2008). Cyclodextrins. *Toxicologic Pathology*, *36*, 30–42.

Tadros, T. F., Izquierdo, P., Esquena, J., & Solans, C. (2004). Formation and stability of nano-emulsions. *Advances in Colloid and Interface Science, 108–109*, 303–318.

Theeuwes, F. (1975). Elementary osmotic pump. *Journal of Pharmaceutical Sciences, 64*(12), 1987–1991.

Theeuwes, F., & Higuchi, T. (1974). ALZA Corp., assignee. Osmotic dispensing device for releasing beneficial agent. US Patent 3,845,770.

Theeuwes, F., Swanson, D. R., Guittard, G., Ayer, A., & Khanna, S. (1985). Osmotic delivery systems for the β-adrenoceptor antagonists metoprolol and oxprenolol: Design and evaluation of systems for once-daily administration. *British Journal of Clinical Pharmacology, 19*(Suppl.), 69S–76S.

Theeuwes, F., & Yum, S. I. (1976). Principles of the design and operation of generic osmotic pumps for the delivery of semisolid or liquid drug formulations. *Annals of Biomedical Engineering, 4*(4), 343–353.

Timmermans, J., & Andre, J. M. (1994). Factors controlling the buoyancy and gastric retention capabilities of floating matrix capsules: New data for reconsidering the controversy. *Journal of Pharmaceutical Sciences, 83*, 18–24.

Timmermans, J., & Moes, A. J. (1990). How well do floating dosage forms float. *International Journal of Pharmaceutics, 62*, 207–216.

Uchegbu, F. I., & Suresh, V. P. (1998). Non-ionic surfactant based vesicles (niosomes) in drug delivery. *International Journal of Pharmaceutics, 172*, 33–70.

Veiga, F. J. B. (1988). Oral medicines of controlled release. *Boletim da Faculdade de Farmacia de Coimbra, 12*(2), 17–87.

Verma, R. K., Arora, S., & Garg, S. (2004). Osmotic pumps in drug delivery. *Critical Reviews in Therapeutic Drug Carrier Systems, 21*(6), 477–520.

Vicentini, F.T.M.C., Depieri, L. V., Polizello, A. N. M., Del Ciampo, J. O., Spadaro, A. C. C., Fantini, M. C. A., et al. (2013). Liquid crystalline phase nanodispersions enable skin delivery of siRNA. *European Journal of Pharmaceutics and Biopharmaceutics, 83*, 16–24.

Villiers, A. (1891). Sur la fermentation de la fecule par l'action du ferment butyrique. *Comptes Rendus de l'Académie des Sciences, 112*, 536–538.

Williams, A. C., & Barry, B. W. (2004). Penetration enhancers. *Advanced Drug Delivery Reviews, 56*, 603–618.

Winsor, P. A. (1948). Hydrotropy, solubilisation and related emulsification processes. *Transactions of the Faraday Society, 44*, 376–398.

Woodley, J. (2001). Bioadhesion, new possibilities for drug administration? *Clinical Pharmacokinetics, 40*(2), 77–84.

Wright, J. C., Leonard, S. T., Stevenson, C. L., Beck, J. C., Chen, G., Jao, R. M., et al. (2001). An in vivo/in vitro comparison with a leuprolide osmotic implant for the treatment of prostate cancer. *Journal of Controlled Release, 75*, 1–10.

Wright, J. C., & Stevenson, C. L. (1999). Pumps/osmotic-introduction. In E. Mathiowitz (Ed.), *Encyclopedia of controlled drug delivery*. New York: John Wiley.

Yuan, J., Liu, T., Li, H., Shi, T., Xu, J., Liu, H., et al. (2014). Oral sustained-release suspension based on a novel taste-masked and mucoadhesive carrier–ion-exchange fiber. *International Journal of Pharmaceutics, 472*, 74–81.

Zhang, J., & Ma, P. X. (2013). Cyclodextrin-based supramolecular systems for drug delivery: Recent progress and future perspective. *Advanced Drug Delivery Reviews, 65*, 1215–1233.

Index

Note: Page numbers followed by *f* indicate figures and *t* indicate tables.

Kolibri 360 GmbH
Ferdinand-Jühlke-Straße 7
99095 Erfurt, Deutschland
produktsicherheit@kolibri360.de